RESEARCH IN MATHEMATICS EDUCATION

VOLUME 3

RESEARCH IN MATHEMATICS EDUCATION VOLUME 3

Papers of the British Society for Research
into Learning Mathematics

Edited by

Candia Morgan and Keith Jones

British Society for Research into
Learning Mathematics

Published and distributed by the British Society for Research
into Learning Mathematics

20 Bedford Way, London, WC1H 0AL, England

Printed and bound by Hobbs the Printers Limited
Brunel Road, Totton, Hampshire SO40 3WX

ISBN 0-9538498-1-3

Cover design: by Simon Rowland

First Published in 2001

CONTENTS

REVIEW PANEL

The editors acknowledge with thanks the contribution of the following members of the *British Society for Research into Learning Mathematics* in the review process.

Janet Ainley

Alan Bell

John Berry

Paul Blanc

Tony Cotton

Janet Duffin

Derek Foxman

Peter Gates

Peter Gill

Simon Goodchild

Linda Haggarty

Tansy Hardy

Tony Harries

Dave Hewitt

Brian Hudson

Barbara Jaworski

Lesley Jones

Steve Lerman

Eric Love

Olwen MacNamara

Sue Martin

Richard Noss

Declan O'Reilly

Sue Pope

Tim Rowland

Ian Stevenson

Rosamund Sutherland

Anne Watson

Julian Williams

Jan Winter

Derek Woodrow

RESEARCH IN MATHEMATICS EDUCATION: SOME ISSUES AND SOME EMERGING INFLUENCES

Keith Jones, University of Southampton

Candia Morgan, Institute of Education, University of London

Research carried out by members of the British Society for Research in Mathematics Education (BSRLM) is not only grounded in knowledge of developments in the field of research and of mathematics classrooms and other sites where mathematics teaching and learning occurs, but is also influenced by developments in education generally (in both practice and policy) and in closely related fields such as psychology and sociology, as well as mathematics itself. Recent research, such as that contained in this volume of BSRLM research papers, has been undertaken against a backdrop of severe criticism of the quality of education research in general, although little of this criticism appears to have been especially directed at mathematics education research. Nevertheless, emerging UK and international developments, some a direct consequence of recent criticisms of educational research, are likely to have a considerable impact on educational research in general, and on research in mathematics education in particular, in the coming years.

The purpose of this chapter is to introduce the papers in this collection of extended research reports from BSRLM [1]. We do that by first reviewing the general criticisms of educational research, looking briefly at the (possible) relationships between research, policy and practice. We then locate this collection of research papers in relation to some of these issues. As research in education is coming under increasing scrutiny it is appropriate to consider the likely impact of emerging research policy developments. These are considered in the second part of this chapter

USES AND ABUSES OF EDUCATIONAL RESEARCH

Users of educational research are, it is claimed, most interested in 'what works'. Griffin (2001), for instance, has observed that the types of 'products' of research in the social sciences that best meet the needs of policy makers are:

- those that provide answers;
- those that provide clear questions in the absence of clear answers;
- those that demonstrate a clear accumulation of findings over time;
- those that can be distilled to a single set of bullets.

Teachers, according to Desforges (2000), want the following from education research:

- standard and stable models of learning;
- coherent, organised, well-established findings;
- vibrant working examples of success;

- research results converted as far as possible into the technologies of education – into curriculum or other pedagogic materials.

Desforges argues that teachers "cannot work with models that change with the wind" and that they "do not have time for literature searches or for refined academic debates". He also suggests that, for teachers, "*That* something works is one thing. Examples of how it can be got to work are crucial." (p.3, emphasis in the original)

Such observations indicate ways in which educational research is said be useful, providing, for example, evidence that can inform both policy and practice. Yet, as Edwards notes:

> Recent calls for education policy and practice to be informed by evidence have been preceded and accompanied by strident complaints about the persistent failure of researchers to provide evidence worth using. (2000, p.3)

Such criticisms, which have been particularly prevalent in the UK, have come from politicians, policymakers and from within the research community itself. Examples include Hargreaves' (1996), assertion that educational research "is poor value for money in terms of improving the quality of education provided in school", and subsequent inquiries funded by the DfEE (Hillage, Pearson, Anderson and Tamkin, 1998) and Ofsted (Tooley and Darby, 1998), both of which were highly critical of educational research in general. A common feature of the criticisms contained in these reports has been their bluntness. Hargreaves, for example, is of the view that there is a considerable amount of

> frankly second-rate educational research which does not make a serious contribution to fundamental theory or knowledge, which is irrelevant to practice; which is unco-ordinated with any preceding or follow-up research; and which clutters up academic journals that virtually nobody reads. (1996, p.7)

The Ofsted-funded review of educational research by Tooley and Darby (1998) is similarly highly critical of educational research in general. Through an examination of a sample of 41 published educational research articles from four prominent journals, Tooley and Darby claim that a majority of educational research does not satisfy criteria of good practice. Most educational research, they declare, is partisan, poorly conducted, and irrelevant. They claim that "the picture emerged of researchers doing their research largely in a vacuum, unnoticed and unheeded by anyone else" (p.6). Even where education research does address policy-relevant and practical issues, Hillage *et al.* (1998) allege that it tends to be:

- small scale and incapable of generating findings that are reliable and generalisable;

- insufficiently based on existing knowledge and therefore capable of advancing understanding;

- presented in a form or medium which is largely inaccessible to a non-academic audience; and

- lacking interpretation for a policy-making or practitioner audience.

Whatever the legitimacy of these general denunciations of educational research – and they have been rigorously contested (see, for example, Hammersley, 1997; Atkinson, 2000; Lagemann, 2000) – the strident criticisms noted above have led to a number of developments in the UK (some of which reflect developments taking place more widely) that will undoubtedly impact on the conduct of research in education in general, and in mathematics education in particular, over the coming years. These emerging developments are examined in the concluding parts of this chapter. For now we turn to the papers in this volume of research reports and examine them in the light of these criticisms of educational research.

RESEARCH PAPERS FROM THE BRITISH SOCIETY FOR RESEARCH IN MATHEMATICS EDUCATION

All the chapters in this volume originate from presentations at meetings of BSRLM during 1999, subsequently published as short papers in the proceedings of the Society. These papers were then substantially revised and expanded for consideration for this more formal publication. Each chapter was subject to blind peer review and only those judged to be of sufficient quality were accepted for publication. The review process, involving a group of experienced members of the Society with a wide range of expertise and interests, provided much constructive feedback, allowing authors to revise and strengthen their chapters. The Society has always sought to encourage and support new researchers as well as to provide a forum for those who are more experienced. This dual aim is reflected in the make-up of the group whose work comprises this volume. The authors range from well-established researchers with strong national and international reputations to new researchers, for whom the chapter included here may be their first refereed publication. They also include teachers of mathematics at various levels and research students as well as those involved in teaching and researching mathematics education in institutions of higher education.

This volume does not attempt to present a complete or representative overview of mathematics education research in Britain. Rather, it provides a snapshot of recent and current work. It displays interest in a broad range of issues in mathematics education, making use of different theoretical frameworks and methodologies and including both reports of empirical studies and more theoretical contributions. The chapters have been organised into four themes:

Mathematics and schooling

Teachers and teacher development

Mathematics, language and meaning

Technology and learning mathematics

Each of these themes is introduced at appropriate points in the volume. The purpose of this introductory chapter is not to describe them in any detail, nor to subject

3

individual chapters to further critical review. Rather, the intention here is to frame the complete set of chapters against some of the issues raised above.

The first thing to observe is that the improvement of the experience of mathematics learners in all contexts is a central concern of each of the chapters in this volume. This applies whether the issue is the appropriateness of early experiences with number provided for young children (Godfrey and Aubrey), the need to make the human face of mathematical writing and mathematical activity more accessible to students at all levels (Morgan), the benefits of linking visual and algebraic ways of thinking in undergraduate mathematics (Chae and Tall), or the different kinds of knowledge that mathematics teacher educators need in order to enable trainee teachers of mathematics to acquire the requisite professional knowledge (Prestage and Perks). This focus on improving the experience of mathematics learners accords with the policy recommendation of the US National Educational Research Policy and Priorities Board that "the priority for research in education must be a high level of achievement for all students" (1999b, p.1), though various parties may well have their own views about what might constitute appropriate definitions of 'high achievement' and how this might be measured.

The concerns of the authors are also in many cases closely related to current priorities for teachers, curriculum developers and policy makers – priorities that may be driven by developments in education policy, by debates about the nature of the mathematics curriculum, by increased availability of new technologies. Thus we see chapters that engage with the teaching and learning of number (Foxman; Godfrey and Aubrey; Bills; Sutherland *et al.*) as the National Numeracy Strategy is introduced by the United Kingdom government. We have investigations into the teaching and learning of proof (Olivero) and algebra (Coles and Brown) – both of which have been identified as areas of weakness within the National Curriculum and in need of development (see Royal Society/JMC, 1996; 2001). National concern with the supply and quality of mathematics teachers and the nature of their induction into the profession is reflected in our choice of *Teachers and teacher development* as one of the four themes in this volume. Similarly, various aspects of the use of new technologies are addressed both in the chapters collected under the *Technology and learning mathematics* theme and elsewhere.

As noted above, Tooley and Darby (1998) generally commend research that builds on what is currently known and utilises a suitable theoretical framework. All the chapters in this volume locate themselves clearly in relation to existing research results and theory; however, 'building on what is known' is not a straightforward exercise. It may involve:

- taking account of results of previous studies when designing new research – as Godfrey and Aubrey have done in order to further illuminate our understanding of early mathematical development;

- addressing questions that have been explicitly or implicitly raised by previous studies – as Smith does in studying the relationships between student teachers' expressed beliefs, their teaching approaches and their experiences during their training;

- extending the application of established theories and the investigation of recognised research issues into new areas of mathematics or different phases of education – as Chae and Tall have done in relation to visualisation and the development of conceptual understanding in the context of undergraduates studying the foundations of chaos theory;

- applying new theoretical insights or methodological tools to existing work – as Foxman has done in his re-analysis of data on mental calculation methods, originally studied in the 1980s;

- developing new theoretical perspectives with which to interrogate or explain previous results or commonly recognised phenomena – as Prestage and Perks and Crisan do in relation to teachers' professional knowledge;

- applying specific theoretical perspectives – as Williams *et al.* do in order to re-conceptualise relationships between mathematics in school or college and in the workplace.

In other cases, however, researchers push the boundaries of our knowledge about mathematics education in rather different ways, including:

- investigating an issue that has previously been neglected – thus Sutherland *et al.* both make an initial contribution to our knowledge about primary mathematics text books and develop a framework for studying them;

- using close analysis of 'natural' or 'experimental' classroom data in order to test out, illustrate and develop theoretical perspectives on mathematics education – thus Coles and Brown address teaching and learning algebra, Olivero investigates students' development of conjecturing and proving processes, and Elliott *et al.* look at the shared construction of meaning in group interaction;

- applying tools from other disciplines to develop new insights – thus Bills infers the nature of children's mental representations of numbers and operations from analysis of their spoken language, using metaphors and other tools from linguistics;

- challenging common assumptions and presenting alternatives – thus Morgan questions perceptions of the nature of mathematical writing, while Huckstep and Rowland argue for caution in invoking the notion of 'creativity' in justification for school mathematics.

The categorisation that we have offered above is, of course, only partial. Most of the chapters may be seen to draw on and build on existing research results, questions and theoretical perspectives in multiple ways.

Given the range of concerns and purposes that we have already indicated, it is to be expected that the methodological approaches vary, since it is, of course, important that the research methods used match the research questions being pursued. Following the categorisation employed by Tooley and Darby (1998, pp.10-11), the majority of the chapters in this volume report studies involving the analysis of empirical data. Several of the chapters are more theoretical in intent, constructing philosophical theses or developing theoretical perspectives, informing their arguments with reference to previous empirical results or to well-chosen examples. The majority of the studies involving analysis of empirical data are what Hillage *et al.* (1998), somewhat disparagingly, call "small-scale", involving what are essentially case studies of single classes, small groups or small numbers of individual students, teachers or texts. Given the limited resources allocated to education research in general, and to mathematics education in particular, it is perhaps inevitable that much research is small-scale. Only two of the studies (Godfrey and Aubrey; Foxman) make use of more substantial data sets (and Foxman is working with data originally collected by the Assessment of Performance Unit nearly fifteen years ago). What is recognised, however, both by Tooley and Darby and by Hillage *et al.*, is that it is not necessarily the size that matters (after all, only one counter-example is needed to disprove a mathematical theorem), but whether a study is capable of helping to generate findings that are reliable and generalisable.

Appropriateness of methodology and generalisability of findings were two important issues addressed by the reviewers of the chapters in this volume. In recommending acceptance for publication, the reviewers judged that the authors undertaking small-scale research were not making unwarranted generalisations but were pointing out how their paper could contribute to the development of more robust findings (something for which Tooley and Darby commend the only mathematics education research paper included in their random selection). Given the range of questions in each of the domains of mathematics education research identified by the CoPrIME/ BSRLM FRAME initiative (Mason, 2001 – discussed below and see Appendix 3), it is perhaps reasonable to conclude that, at the current level of funding, this represents a commendable achievement by these BSRLM researchers.

The purpose of this introductory chapter is not to subject the papers in this volume to the level of analysis undertaken by Tooley and Darby (1998). In their identification of themes such as partisanship (in terms, for instance, of the conduct and presentation of educational research), methodological weaknesses, and relevancy to practice, Tooley and Darby provide interested readers with possible means of engaging with the authors of the papers, or perhaps directly with the Executive Committee of BSRLM who oversee the production of volumes in this series. Certainly, the reviewers of individual papers, and the editors who oversaw the revision of the papers, paid special attention to the quality of the design, conduct and reporting of each research study published in this volume.

Hopefully, the reader will find, if not answers, then clear questions that can help to lead to a clear accumulation of findings over time (even if these may not be able to be distilled to the single set of bullet points apparently beloved by policy makers). A particular feature of this collection is the focus on questions that are undoubtedly important to classroom practitioners. These include, for example, children's mental mathematics strategies (Foxman), the relationship between school-taught mathematics and that used in the workplace (William *et al.*), how mathematics teachers learn to use ICT in their teaching (Crisan), the relationship between trainee teachers' beliefs and their teaching approach (Smith), the possible impact on pupils' learning of representations in mathematics textbooks (Sutherland *et al.*), and the use of dynamic geometry software for conjecturing and proving in geometry (Olivero). While this collection of chapters may not provide teachers with everything they are looking for from research, it does contain working examples and, in some cases, findings that may contribute to the development and "warranting" (Ruthven, 1999) of suitable curriculum and other pedagogical materials and practices.

While considering the relevance of research to communities of users, it is necessary to address its accessibility (or inaccessibility) to members of these communities. The writing in this volume, arising as it does from meetings of BSRLM, is primarily addressed to an audience similar to that from which the membership of BSRLM is drawn. In other words, we expect the chapters to be most accessible to researchers and students in mathematics education and to those teachers, curriculum developers and policy makers who are already active readers and users of research. We hope that the writing is not unnecessarily obscure, but we recognise that it is not primarily addressed to practitioners. Having said this, many of the authors whose research is represented here also report their work directly to practitioners in professional journals such as *Mathematics Teaching* and *Mathematics in School*. Recently, abstracts of the research reported at BSRLM meetings have also started to be included in these journals. The joint FRAME initiative undertaken by CoPrIME and BSRLM (Mason, 2001) is similarly intended to address an audience of policy makers. By these means BSRLM is taking steps to widen the dissemination of the useful findings of research in mathematics education, to encourage dialogue between researchers, practitioners and policy makers and, by raising awareness of the activities of members of BSRLM, to widen participation in and with research among practitioners themselves.

While the research reported in this volume was undertaken against a backdrop of severe criticism of educational research in general, future research is going to be influenced by emerging trends in education research policy. The rest of this chapter reviews these emerging trends, beginning with a brief commentary on the current state (and status) of research in mathematics education.

THE STATE OF RESEARCH IN MATHEMATICS EDUCATION

One positive indicator of the quality of work in mathematics education comes from an unlikely source. As noted above, the Ofsted-funded review of educational research (Tooley and Darby, 1998) is highly critical of educational research in general, claiming that a majority of such research does not satisfy criteria of good practice. By chance, one of the articles analysed by Tooley and Darby is from the mathematics education community (the article is Coe and Ruthven, 1994). In their report, Tooley and Darby highlight the article as an *example of good practice in methodology*. Tooley and Darby note that in the Coe and Ruthven article:

> Background ideas on the importance of proof in mathematics are clearly and comprehensively explored. The research is a small-scale qualitative study, but the authors do not make unwarranted generalisations from it. They treat it in the vein of a pilot study, pointing out how the methodology could be used for a larger study from which generalisations could be made. (1998, p.47)

Further evidence of the status of research in mathematics education comes from the assessment of the quality of UK education research by Professor Michael Bassey, Executive Secretary of the British Educational Research Association (BERA) and not himself a mathematics educator. Writing in the Times Higher Education Supplement of Friday 4 December 1998, Bassey suggests that the following categories of UK education research are 'world class' quality:

- school improvement;
- cognitive acceleration in science education;
- mathematical education;
- formative assessment;
- school action research;
- gender awareness.

Another example of the importance of UK research in mathematics education is that BERA have included it in their recent initiative to provide a 'map' of educational research in the UK. A major part of this initiative has been a series of 13 national events, each covering a different aspect of educational research, leading to a series of monographs. One of these events surveyed UK research in mathematics education, focusing on the teaching and learning of number, and was organised in collaboration with BSRLM. The monograph containing this research review is due to be published in 2001 by BERA and BSRLM (Brown and Askew, in preparation).

Not only are there these positive signs of the quality of research in mathematics education in the UK, a substantial quantity is also published. One indication of the volume of UK research in mathematics education can be found in the analysis of educational research submitted for the 1996 research assessment exercise (RAE) (Kerr, 1998). A major component of the RAE uses the details of the best four research publications published over the previous four years by each active UK researcher employed by a higher education institution. The approach adopted in Kerr's analysis of educational research was to identify key themes in the research and to map the concentration of research effort within and across those themes. Publications could be counted in more than one category. This approach found the largest number of publications to be in education policy (47% of all papers), with 'subject-based enquiry' second (31.1% of all papers). Of the 3115 papers classified as 'subject-based enquiry', the first three categories were:

Science education   513

Mathematics education   419

English education   268

This suggests that mathematics education is the second largest area of subject-based research in curriculum areas in the UK.

While there are thus some positive signs about research in mathematics education, we have to look to the future and, in particular, at emerging policy developments that are likely to impact on such work. These are examined in the final sections of this chapter.

SOME DEVELOPMENTS IN POLICY ON EDUCATIONAL RESEARCH

Significant recent developments in policy on educational research include:

- the establishment in the UK in September 1999 of a National Education Research Forum (NERF) and the beginnings of the development of a national education research framework (see http://www.nerf-uk.org/);

- the establishment internationally in 1999 of the 'Campbell Collaboration', an emerging international effort that aims "to help people make well-informed decisions by preparing, maintaining, and promoting access to systematic reviews of studies on the effects of social and educational policies and practices" (see http://campbell.gse.upenn.edu/).

While each of these developments is treated in more detail in Appendices 1 and 2 respectively, it is worth noting here that the main aim of NERF is to "develop a strategy for educational research, shape its direction, guide the coordination of its support and conduct, and promote its practical application". To some extent, this mirrors similar developments in other countries. In the USA, for example, a National Educational Research Policy and Priorities Board (NERPPB) was established in 1994 "to forge a national consensus with respect to a long-term agenda for educational research, development, dissemination" (NERPPB, 1999a). The NERPPB also recognises that such research requires funding, arguing that "funding for educational research must be increased dramatically" and giving an interim target of about $1.5 billion annually.

The work of NERF has not yet reached the level of detail of the NERPPB. It has neither established specific priorities for educational research, nor explicitly stated

9

that an increase in funding for educational research, from what it identifies as less than 0.5% of the total national expenditure on education, is required in the UK. Yet developments internationally, described in more detail in Appendix 2 but summarised below, may well have a major impact on its work.

A particularly pertinent international development is the Campbell Collaboration. This enterprise intends to be a "multi-national collaboration" producing "systematic reviews" of which innovations in education (and in the social and behavioural sectors more widely) "work and fail". The Campbell Collaboration has already formed an Education Co-ordinating Group [2], produced draft review protocols (ways of generating reviews) and appointed lead reviewers for reviews of truancy programs, voluntary tutoring, peer assisted learning, and second language training. Its latest report (Campbell Collaboration Secretariat, 2001) indicates that negotiations have been undertaken to determine the feasibility of reviews on teacher induction and mentoring and on monetary incentives and educational achievement. In addition, lead persons have been identified for developing review protocols in mathematics learning, science learning, work-related learning and transferable skills, medical education, leadership and management in education, and assessment and learning.

In terms of mathematics learning the Campbell Collaboration's *Organisational Framework for Preparing and Maintaining Systematic Reviews in Education* (see Davies, Wolf and Holmes, 2001) so far identifies some eight key papers in mathematics education research; the oldest being Hartley's meta-analysis of the effects of individually-paced instruction in mathematics (Hartley, 1977), and the most recent being Hembree's meta-analysis of the nature, effects, and relief of mathematics anxiety (Hembree, 1990). At the time of writing, the Education Coordinating Group is seeking a lead person for the mathematics and science reviews, although some preparation work has already been done (see Leow and Boruch, 2001).

While these two developments (NERF and the Campbell Collaboration) have yet to have a major impact on UK research in mathematics education, the need to review what is known through research, reflected in such developments, has not gone unnoticed in the international mathematics education research community. This has already led to some developments in policy on research in mathematics education.

DEVELOPMENTS IN POLICY ON RESEARCH IN MATHEMATICS EDUCATION

The 1990s can be characterised as a time when research in mathematics education began to take stock. Examples of substantial efforts to review and define the field include the NCTM's handbook of research in mathematics education (Grouws, 1992), the ICMI study of the nature and results of mathematics education research in 1994 (and the resulting two volumes published subsequently, see Sierpinska and Kilpatrick, 1998), and encyclopaedias from publishers including Kluwer (Bishop, Clements, Keitel, Kilpatrick and Laborde, 1996), RoutledgeFalmer (Grinstein and

Lipsey, 2001), and, shortly, LEA (English, Bartolini Bussi, Jones, Lesh and Tirosh, in preparation). A review of research published in the BSRLM proceedings is currently underway (Nickson, in preparation).

Published reviews of these efforts to variously summarise and evaluate the considerable amount of research in mathematics education have generally been favourable. Nevertheless, not all found that these publications placed mathematics education research in a positive light. For example Steen (1999, p236), a prominent mathematics educator, thought that the ICMI study volumes "document a field in disarray, a field whose high hopes for a science of education have been overwhelmed by complexity and drowned in a sea of competing theories". If this is the case then it seems that further efforts by the mathematics education community would appear to be needed to achieve what Lagermann (2000, p. ix) calls "a high degree of internal coherence" necessary for sustained progress.

In the UK, some effort in this direction can perhaps be seen in the ongoing project initiated by the Committee of Professors in Mathematics Education (CoPrIME) and BSRLM and entitled the Formulation of a Research Agenda in Mathematics Education (FRAME) (see Mason, 2001). As a result of the work done so far on this initiative, the framework consists of nine domains of research (see Appendix 3), which represent a distillation of issues that CoPrIME and BSRLM have identified. Each domain of enquiry attempts to capture what the authors believe "need to be researched carefully in order to provide evidence which can inform both policy and practice". An extended account of domain 5, on Mathematics Teachers' Professional Education and Development, is shortly to be published by BSRLM (see McNamara, Jaworski, Rowland, Hodgen, Prestage and Brown, in preparation)

This CoPrIME/BSRLM initiative demonstrates that the accumulation of evidence from research in mathematics education has made it possible to identify clear questions that warrant further research. The next section considers emerging influences on how such research may be carried out.

EMERGING DIRECTIONS FOR THE FORM OF EDUCATION RESEARCH

Towards the end of their review of the evidence base for what they call effective teaching, Muijs and Reynolds comment that, "there are robust findings on the learning and teaching of reading and literacy, and an increasingly strong research base of findings on the learning and teaching of mathematics (although this research base remains less strong than that on teaching literacy)" (2001, p.211). Precisely what Muijs and Reynolds mean by 'robust' is never made entirely clear but it may be assumed to be about how valid, reliable and generalisable research results are. It seems likely to be particularly concerned with the research design, including the use of comparison groups, representative samples, and the degree of replication (so that the intervention can be judged in various contexts).

11

Generalisable results tend to come from accumulated evidence from a series of smaller-scale projects, exemplified in the research presented in this volume, or from very large-scale projects. As Tooley and Darby (1998) observe (without seeming to notice that it is related to their criticism of the preponderance of small-scale research) there has been, and is, insufficient large-scale research in education in the UK.

While NERF has yet to say much if anything about the relative merits of funding many small projects against funding a smaller number of larger projects (assuming a fixed level of funding), one current example of funding a few large-scale projects is the ESRC teaching and learning research programme. Despite a budget of £23 million, the objectives of the Programme are extraordinarily ambitious and include "enhancing the achievement of learners at all ages and stages in education, training and life-long learning" (see http://www.ex.ac.uk/ESRC-TLRP/). Of the four Research Networks funded in Phase I of this initiative, and the nine Research Projects being funded under Phase II, there is only one that makes explicit reference in its title to an aspect of mathematics education (*The Role of Awareness in the Teaching and Learning of Literacy and Numeracy in Key Stage 2* directed by Terezinha Nunes, Peter Bryant and Jane Hurry) although it is likely that other TLRP projects have mathematics education elements.

This brings the choice into sharp relief. If funding for educational research in the UK is not raised to a level comparable with that in the USA (where research grants of over a $1 million are not unusual) the education community in the UK will have to continue to rely on the accumulation of evidence from smaller-scale studies or make do with a far smaller number of relatively large-scale studies.

While funding considerations may well move towards funding a relatively small number of reasonably large-scale studies, there are some strong forces coming to bear on the form this research may take. For example, the Campbell Collaboration Education Co-ordinating Group, in its plan to publish ten to fifteen systematic reviews each year (Campbell Collaboration Secretariat, 2001), is to focus almost exclusively on the results of research that employs a randomised control trial methodology (together with the use of previously published systematic reviews and meta-analyses such as those noted above by Hartley, 1977, and Hembree, 1990). According to the current Campbell Collaboration Secretariat report, the Campbell Controlled Trials Register (C2-SPECTR) already has over 10000 entries covering Social, Psychological, Educational, and Criminological research.

Another example is the review by Dixon, Carmine, Lee and Wallin (1998) of what they call "high quality experimental mathematics [education] research". In this somewhat contested review (see, for example, Kilpatrick, 1999; Becker and Jacob, 2000), the authors argued that of some 8727 studies in mathematics education published since 1970, only 956 articles satisfied "minimum identification criterion of being an experimental study of mathematics" (p.3). Furthermore, of the 956 studies, only 110 met further criteria (to do with an interpretation of various aspects of validity) imposed by Dixon *et al.* In the end, the review consisted of the findings from just these 110 studies.

While this is not the place to discuss in detail the contribution of randomised control trial (or experimental) methodology, it is worth noting that exclusive reliance on their use is not uncontested, even in medical research, sometimes invoked as a model to which educational research might aspire (see Howie, 2000; Pirrie, 2001). There is a danger that forms of knowledge that may be equally useful, gained by other methodologies, will be devalued and neglected. As Kilpatrick (1999) notes, in terms of research in education, this restriction to experimental studies, selected according to criteria such as those imposed by Dixon *et al.* or by the Campbell Collaboration, represents a common but unfortunate misunderstanding of educational research and the kind of knowledge it can contribute to efforts to improve teaching and learning. First, educational methods that people are interested in implementing do not have the specificity of medical treatments. Secondly, research cannot on its own decide what happens in places where mathematics teaching and learning occurs because it cannot be used to resolve matters that are really about values and priorities. Thirdly, teachers have a critical role in influencing how mathematics teaching and learning occurs. In medicine, drugs and therapies depend much less on the competence and attitude of the administering physician than teaching does on the competence and attitude of the teacher.

CONCLUDING COMMENTS

At the moment it appears that NERF has run out of steam. Apart from recently publishing an analysis of the responses they received to their consultation paper on 'Research and Development for Education' (NERF, 2001) NERF appears to have no specific objectives, targets or timeframe. How they will produce a national education research framework for UK educational research is anyone's guess. As their consultation response paper records "whilst the idea of some sort of framework for educational research gains support there is enormous reservation about the Forum and the strategy in terms of aims, scope and ability to deliver" (p.2).

NERF could do well to examine a recent analysis of the impact of educational research in Australia (particular educational research carried out in Australia) as the report claims to provide "compelling evidence that Australian educational research is respected internationally and makes a difference in the worlds of schools, and policy development" (Department of Education, Training and Youth Affairs, 2000, p.4). For example, a survey carried out for the inquiry found that "almost all the school principals, professional associations of educators, and school system administrators [sampled in Australia] expressed the view that educational research had benefited Australian education" (p.5). In the UK, a recent paper by Galton (2000) reports evidence that teachers in the UK take the results of educational research seriously.

In the US, the NERPPB has, since 1994, undertaken systematic investigation on the dimensions and scope of educational research and development and, based on this work, has published a first comprehensive statement on research in education (see NERPPB, 1999b). This document gives the first goal for education research as raising student achievement, stating that "the priority for research in education must be a high level of achievement for all students, and, within that domain, the initial emphasis should be on *reading* and *mathematics* achievement" (emphasis in original) (p.1). In mathematics, the document states, research is needed on "why students have so much trouble making transitions (e.g., from concrete objects to more abstract ideas), understanding formal representations, multiplicative reasoning, and essential mathematical and statistical concepts, such as chance, randomness, and probability."

Perhaps NERF will take notice of this US report and of the CoPRIME/BSRLM initiative to develop a research agenda for UK research in mathematics education. Or perhaps UK involvement at Government level in the Campbell Collaboration, which is already high, will take things in another direction.

According to the Campbell Collaboration Secretariat report, Philip Davies, Co-chair of the Campbell Collaboration Education Coordinating Group is Deputy Director of Policy Evaluation and Policy Studies in the UK Cabinet Office (as well as professor in the Department for Continuing Education at the University of Oxford). In a presentation at the BERA conference in 1999, Judy Sebba, the main Campbell Collaboration contact at the UK Department for Education and Skills (DfES), explained UK Government thinking on developing evidence-informed policy and practice in education (Sebba, 1999). In her talk she offered her 'litmus test' for this initiative:

> In ten years time a pupil will be able to access a website to find out the best way to be taught fractions. If the teacher is not using that method the pupil can ask why not

This echoes both the 'consumer network' of the Cochrane collaboration (see Appendix 2), where anyone can get evidence to help them "make decisions about health care", and the people from NICE (The National Institute for Clinical Excellence), who provide "authoritative, robust and reliable guidance on current best practice". When asked, at her BERA presentation, what would happen if the teacher in her example was using a superior method to that given on the website, Judy Sebba appeared unable to provide an adequate response. While there are another eight years left to find out what will become of Sebba's litmus test, it does focus attention on just what are reasonable expectations of research in mathematics education.

NOTES

1. All the BSRLM papers referred to in this chapter appear in C. Morgan and K. Jones (eds), *Research Papers in Mathematics Education, volume 3*. London: BSRLM.

2. The Campbell Collaboration Education Group has a website but at the time of writing it has no content. See: http://aix1.uottawa.ca/~iupsys/campbell/education.html

All URLs quoted in this chapter were in operation when this chapter went to print (September 2001).

APPENDIX 1

THE UK NATIONAL EDUCATION RESEARCH FORUM (NERF)

Following its establishment in 1999, the first task NERF has undertaken has been to produce discussion documents (available from NERF) on each of the following:

- identifying priorities for educational research;
- the quality of educational research;
- building research capacity;
- research funding;
- the impact of research on policy and practice.

The key recommendations in these discussion documents are:

- investment in an ongoing horizon-scanning exercise, designed and executed specifically for education as a method of identifying educational research priorities;
- that, within its own terms, any particular research study or project should be of good quality with respect to those terms;
- a systematic appraisal of the current state of research capacity – perhaps initially through a consultation process informed by a commissioned review drawing together and supplementing existing reports, but eventually through the maintenance of a 'register' of capacity;
- the establishment of a funders' group to enable funders of educational research to share thinking about priorities and to exchange information about practices;
- the development of a model of educational research and the place of impact describing the means by which research can have a warranted impact on policy and practice.

In addition to the establishment of NERF, several research centres have been set up: the Centre for Economics of Education; the Centre for ICT in Education; the Wider Benefits of Learning Research Centre; and the Evidence for Policy and Practice Information and Co-ordinating Centre (EPPI-Centre).

Alongside these developments are continuing efforts to involve teachers in schools in research (through Government initiatives such as 'Professional Bursaries', 'Best Practice Research Scholarships', and research opportunities through the Teachers' International Professional Development programme). In addition, a database called Current Educational Research in the UK (CERUK,) and maintained by the National Foundation for Educational Research (NFER), has been established to record all UK educational research at PhD level and above. The main UK funding body for educational research, the Economic and Social Research Council (ESRC) has also set up an Evidence-Based Policy and Practice Network (not restricted to education) to "improve the capacity for exchange of research-based evidence between public policy researchers and practitioners and to contribute to the improvement of quality of research, policy development and practice" (see http://www.evidencenetwork. org/).

APPENDIX 2

THE CAMPBELL COLLABORATION

The Campbell Collaboration, established in 1999, is named after the American psychologist Donald Campbell (1917 –1996) who consistently drew attention to what he saw as the need to assess more rigorously the effects of research efforts, especially in the social sciences. Boruch, Petrosino and Chalmers (1999) provide three reasons for establishing the organisation: (a) a surge of interest in high-quality evidence for public policy decisions, (b) the increased frequency in using randomised field trials to discern the effects of new programs or new variations on existing programs, and (c) the increased frequency of systematic reviews.

The precedent for the Campbell Collaboration is the Cochrane Collaboration in health care (see http://www.cochrane.org), which was established in 1993. The Cochrane Collaboration has, to date, produced over 1000 systematic reviews of studies of health-related interventions, and has over 800 further reviews in preparation. They have a 'consumer network' (see http://www.cochraneconsumer.com) where anyone can get summaries of Cochrane reviews of evidence to help them "make decisions about health care". In a related development in the UK, NICE – The National Institute for Clinical Excellence – was set up as a Special Health Authority for England and Wales on 1 April 1999. It is part of the National Health Service (NHS), and its role is to provide patients, health professionals and the public with "authoritative, robust and reliable guidance on current best practice" (see http://www.nice.org.uk). The Campbell Collaboration aims to build on the experience of the Cochrane Collaboration and, presumably, develop the same form of products.

APPENDIX 3

COPRIME/BSRLM FORMULATION OF A RESEARCH AGENDA IN MATHEMATICS EDUCATION (FRAME)

Currently, the framework consists of nine domains of research (see Mason, 2001).

Domain 1: Curriculum Design and Implications

Who decides the intended mathematics curriculum, and what are their intentions? How do these intentions manifest themselves in the curriculum as planned, as taught, and as experienced by children? What is the background philosophy and what are the underlying assumptions being made by the current National Curriculum and the National Numeracy Strategy? Given the form and format of the National Curriculum and the National Numeracy Strategy, what perception of mathematics and of mathematics education (teaching and learning) is being promulgated as a result? What are the social and educational implications of this perception?

Domain 2: Digital technologies

What epistemological and pedagogical criteria drive the design and use of digital technology in support of mathematics education? What are the implications of restricted access to technology through policy and through economic inequities, as experienced both here and abroad?

Domain 3: Levels

What are the criteria which determine achievement at level 4 in the mathematics National Curriculum? Are these criteria adequately tested by current tests? Are the test results robust against changes in time of year, changes in questions, and are they robust over time? How are the criteria themselves changing over time? Do they really represent what we want pupils to achieve?

Domain 4: Mathematical Topics and Themes

In order to inform teacher practices in specific topics, we need to know a great deal more about the struggles and difficulties which pupils experience when teachers use specific approaches in specific topics in mathematics.

Domain 5: Professional development

Numerous training programmes for mathematics teachers are in operation, or have been recently, with many variants (different ITE routes, 20 day courses, lead teachers, ICT training for all teachers, NNS cascade training, needs assessment for KS2, updating of teachers at KS3, Thinking Skills, ...). What impact are these having on their intended audiences, and is there visible evidence of positive change in pupils' experience as a result?

Domain 6: Preserving Strengths

Is the weakened emphasis on practical work and problem solving aspects in the current national policy in mathematics building on strengths, and is it having an effect on pupils' performance on TIMMS-like tests? How would a sample of UK students score now?

Domain 7: Mathematics in the Curriculum

What aspects of mathematics and of mathematical thinking are essential for full participation in a democratic society? What are the different trajectories of children coming from different backgrounds (socio-economic, gender, ethnic origin, ...) through public examinations and tests in mathematics, and through post-compulsory education?

Domain 8: Top-Down and Bottom-Up

How is it possible to balance standardised curricula with individual needs and differences? How can a teacher of thirty or more pupils in each lesson address individual needs, and how do national policies and espoused practices support teachers in this?

Domain 9: How Mathematics is Perceived

Is there evidence that attitudes and beliefs about mathematics are becoming more positive as a result of current initiatives? Is mathematics seen by pupils as opening up their future options?

REFERENCES

Atkinson, E.: 2000, 'In defence of ideas, or why 'What works' is not enough.' *British Journal of the Sociology of Education, 21*(3), 317-330.

Becker, J. and Jacob, B.: 2000, 'The politics of California school mathematics: the anti-reform of 1997-99.' *Phi Delta Kappan, 81*(7), 529-537.

Bishop, A. J., Clements, K. Keitel, C. Kilpatrick, J. and Laborde, C. (eds.): 1996, *International Handbook of Mathematics Education*. Dordrecht: Kluwer Academic.

Boruch, R., Petrosino, A. and Chalmers, I.: 1999, *The Campbell Collaboration: a proposal for systematic, multi-national, and continuous reviews of evidence.* Background paper for the meeting at The School of Public Policy, University College London, 15/16 July 1999.

Brown, M and Askew, M.: in preparation, *Teaching and Learning Numeracy: Policy, Practice and Effectiveness*. Nottingham: BERA/BSRLM.

Campbell Collaboration Secretariat: 2001, *The Campbell Collaboration: Concept, Status, and Plans*. Philadelphia PA: Campbell Collaboration.

Coe, R. and Ruthven, K.: 1994, 'Proof practices and constructs of advanced mathematics students.' *British Educational Research Journal, 20*(1), 41-53.

Davies, D. Wolf, F. M. and Holmes, L.: 2001, *An Organisational Framework for Preparing and Maintaining Systematic Reviews in Education: A Discussion Document*. University of Pennsylvania, Philadelphia PA: Campbell Collaboration Secretariat.

Department of Education, Training and Youth Affairs: 2000, *The Impact of Educational Research*. Canberra, ACT: DETYA.

Desforges, C. W.: 2000, *Familiar Challenges and New Approaches: necessary advances in theory and methods in research on teaching and learning*. The Desmond Nuttall/Carfax Memorial Lecture, BERA, Cardiff, 2000.

Dixon, R. C., Carnine, D. W., Lee, D.-S., and Wallin, J.: 1998, *Review of High Quality Experimental Mathematics [Education] Research: Report to the California State Board of Education and addendum to principal report*. Eugene: University of Oregon, National Center to Improve the Tools of Educators.

Edwards, T.: 2000, *Some Reasonable Expectations of Educational Research*. UCET Research Paper No 2.

English, L., Bartolini Bussi, M. G., Jones, G., Lesh, R., and Tirosh, D. (eds.) (in preparation), *Handbook of International Research in Mathematics Education*. Mahwah, NJ: Lawrence Erbaum Associates.

Galton, M.: 2000, *Integrating Theory and Practice: Teachers' Perspectives on Educational Research*. Paper presented at the Teaching and Learning Research Programme Conference, 9 -10 November 2000, Leicester.

Griffin, J.: 2001, *Who Are the Users?* Paper presented at The Campbell Collaboration First Annual Colloquium, February 2001, University of Pennsylvania, Philadelphia, PA.

Grinstein, L. S. and Lipsey, S. I. (eds.): 2001, *Encyclopedia of Mathematics Education*, New York: RoutledgeFalmer.

Grouws, D. A. (ed.): 1992, *Handbook of Research on Mathematics Teaching and Learning*. New York: Macmillan.

Hargreaves, D. H.: 1996, *Teaching as a Research-based Profession: possibilities and prospects*. The Teacher Training Agency Annual Lecture 1996.

Hartley, S. S.: 1977, *Meta-analysis of Effects of Individually Paced Instruction in Mathematics*. Doctoral Dissertation, University of Colorado.

Hammersley, M.: 1997, 'Educational research and teaching: a response to David Hargreaves' TTA lecture.' *British Educational Research Journal, 23*(2), 141- 161.

Hembree, R.: 1990, 'The nature, effects, and relief of mathematics anxiety.' *Journal for Research in Mathematics Education, 21*, 33-46.

Hillage, J., Pearson, R., Anderson, A. and Tamkin, P.: 1998, *Excellence in Research on Schools*. The Institute of Employment Studies. London: DfEE Publications.

Howie, J. G. R.: 2000: 'Is it knowing or understanding that matters?' *Medical Education, 34*, 246-247.

Kerr, D.: 1998, *Mapping Educational Research in England: An analysis of the 1996 Research Assessment Exercise*. Bristol: National Foundation for Educational Research and the Higher Education Funding Council for England

Kilpatrick, J.: 1999; *The Role of Research in Improving School Mathematics*. Invited Address to the AMS-MAA-SIAM Joint Mathematics Meetings, San Antonio, Texas, January 13-16, 1999.

Lagemann, E. C.: 2000, *An Elusive Science: The Troubling History of Education Research*. Chicago: University of Chicago Press.

Leow, C. and Boruch, R. F.: 2001, *Locating Randomized Experiments on Math and Science Education: A Hand Search and Machine-Based Searches of the American Educational Research Journal*. University of Pennsylvania, Philadelphia PA: Campbell Collaboration Secretariat.

Mason, J.: 2001, *FRAME: Formulating A Research Agenda for Mathematics Education*. Milton Keynes: CoPrIME.

McNamara, O., Jaworski, J., Rowland, T., Hodgen, J., Prestage, S. and Brown, T.: in preparation, *Mathematics Teaching and Teachers' Professional Education and Development: Formulating a Research Agenda for Mathematics Education*. London: BSRLM.

19

Muijs, R.D., Reynolds, D. (2001). *Effective Teaching: Evidence and Practice.* London: Paul Chapman Publishing.

National Educational Research Forum: 2001, *Analysis of responses to 'Research and Development for Education: A national strategy consultation paper'.* London: NERF.

National Educational Research Policy and Priorities Board: 1999a, *The National Educational Research Policy and Priorities Board: Its Role, Development and Prospects.* Washington, DC: NERPPB.

National Educational Research Policy and Priorities Board: 1999b, *Investing in Learning: A Policy Statement with Recommendations on Research in Education by the National Educational Research Policy and Priorities Board,* Washington, DC: NERPPB.

Nickson, M.: in preparation, *A Review of BSRLM Research, 1995-2000.* London: BSRLM.

Pirrie, A.: 2001, 'Evidence-based practice in education: the best medicine?' *British Journal of Educational Studies, 49*(2), 124-136.

Royal Society/JMC: 1996, *Teaching and Learning Algebra pre-19.* Report of a Royal Society/JMC working group chaired by Professor R. Sutherland. London: The Royal Society.

Royal Society/JMC: 2001, *Teaching and Learning Geometry 11-19.* Report of a Royal Society/JMC working group chaired by Professor A. Oldknow. London: The Royal Society.

Ruthven, K.: 1999. 'Reconstructing professional judgement in mathematics education: From good practice to warranted practice.' In C. Hoyles, C. Morgan and G. Woodhouse (eds.), *Rethinking the Mathematics Curriculum* (pp. 203-216). London: Falmer.

Sebba, J.: 1999, *Developing evidence-informed policy and practice in education.* Paper presented at the British Educational Research Association Conference, University of Sussex, Brighton, 2-5 September, 1999.

Sierpinska, A. and Kilpatrick, J. (eds.): 1998, *Mathematics Education as a Research Domain: A Search for Identity.* Dordrecht: Kluwer

Steen, L.A.: 1999, 'Theories that gyre and gimble in the wabe.' *Journal of Research in Mathematics Education, 30*(2), 235-241.

Tooley, J. and Darby, D.: 1998, *Educational Research: a critique.* London: Office for Standards in Education.

THEME 1
MATHEMATICS AND SCHOOLING

The chapters in this section range across sectors of education, from the early years of primary school to the 16-19 age group, moving from school and college into the workplace. The authors also address a wide range of mathematical topics. What all the chapters have in common, however, is, on the one hand, detailed concern with the mathematical thinking of the children and young adults involved and, on the other hand, thoughtful consideration of the ways in which this thinking and its development may be related to teaching approaches, curricula, and the context of learning.

Amid international concern to raise standards of numeracy among children and among the population as a whole, there is a tendency in England to assume that more arithmetic should be taught to more children, starting at an ever earlier age. Ray Godfrey and Carol Aubrey present some results of an international study that cause them to question this assumption. Their multilevel analysis of test data for children in their first year of schooling suggests that, though English children make progress on arithmetic tasks, they make less substantial progress on logical tasks that may underpin further mathematical development. The authors suggest that early emphasis on formal arithmetic may not be productive in the long run.

Mental calculation methods play a major part in current discussions of primary school mathematics in the UK and recent curriculum developments have placed much emphasis on teaching such methods. Derek Foxman's chapter provides a useful reminder that many children develop their own methods without explicit teaching. His re-analysis of data from the surveys of 11- and 15-year olds carried out by the Assessment of Performance Unit in 1987 identifies the methods chosen by children at a time when mental calculation methods were unlikely to be taught in school. He also analyses the relative success of children using different methods. His conclusions suggest some immediate implications for current teaching as well as questions for further research.

Algebra in the early years of the secondary school is perhaps the area of the curriculum that is most guilty of the Cockcroft Committee's complaint that "mathematics lessons are very often not about anything". Alf Coles and Laurinda Brown have set themselves the task of making algebraic representation and activity meaningful by developing a classroom culture in which students find a need to use algebraic symbolism. They examine the teacher's and students' use of 'meta-commenting' on mathematical behaviours and, using a case study of one student's developing use of algebra, propose that a need for algebra may arise as students ask their own 'meta-level' questions.

Employers' complaints that young people lack the mathematical knowledge and skills needed in the workplace are often blamed on inadequacies in the curriculum or, from a rather different point of view, on the problem of 'transfer' of knowledge from one situation into another. Julian Williams, Geoff Wake and Nick Boreham present an alternative perspective that identifies differences in the goals and structures of the activities involving mathematics in school or college and in the workplace. They develop a theoretical description of these differences through analysis of a case study of a successful college student facing difficulties in making sense of graphical data encountered in her workplace experience.

NEEDING TO USE ALGEBRA

Alf Coles, Kingsfield School, South Gloucestershire

Laurinda Brown, University of Bristol, Graduate School of Education

We have completed a Teacher Training Agency (TTA) funded project looking at the teaching and learning of algebra with one mixed- ability year 7 class (11-12 year-olds in the UK). We take Kieran's (quoted in Sutherland, 1997) definition of algebra and see ourselves as developing a 'community of practice' (Lave and Wenger, 1991) in the classroom where the practice is not that of mathematician but of inquirer (Schoenfeld, 1996) into mathematics. The teacher in this community acts as role model for inquirer and metacomments (Bateson, 1972) on the practice of inquiry. In this paper we present evidence for how such metacommenting supports the development of a 'community of inquiry' which leads to the 'need' for algebraic activity. We conclude with a case study, illustrating one student's need for algebra.

BACKGROUND

Issues relating to algebra have formed conclusions to two recent reports (Winter, Brown and Sutherland, 1997, Sutherland, 1997) into mathematics teaching and learning at secondary schools in the UK. Winter *et al.* (1997) is a national report in which algebra is identified as a key component in facilitating a smooth transition for students between school and higher education in mathematics. Students' algebraic skills were often found to be in need of attention by teachers at the start of higher education courses. The Sutherland (1997) report was the outcome of a Royal Society (RS) and Joint Mathematical Council (JMC) working group, set up in 1995 to make recommendations about the teaching of algebra, partly in response to the apparent lack of articulation between mathematics taught at school and that required by higher education (p. ii).

In the RS/JMC report a key conclusion is that:

> the National Curriculum is currently too unspecific and lacks substance in relation to algebra. The algebra component needs to be expanded and elucidated – indeed rethought (p. iii).

A further conclusion is: "that more research is needed to understand the relationship between what algebra is taught and what is learned" (p. iii).

Both these conclusions sparked our interest in looking at algebra in secondary schools. In particular, we wanted to work on Sutherland's (1991) challenge:

> Can we develop a school algebra culture in which pupils find a need for algebraic symbolism to express and explore their mathematical ideas? (p.46)

We have been exploring, through a Teacher Training Agency (TTA) funded project, the developing algebraic activity of one year 7 mixed ability group in a comprehensive school: Alf is the classroom teacher and Laurinda a visiting teacher-researcher.

WHAT IS ALGEBRA?

In asking ourselves the question: "What is algebra?" we wanted to find a definition with which we could work to try to understand what students actually did which could be described as algebraic activity when they were engaged in doing mathematics. In reviewing current research on algebra, strands emerged to do with context, meaning-making, complexity and control which we found useful in our thinking. Introducing and using algebra in a context is talked about from a view, which we support, that:

> Traditionally, algebra in schools has been dealt with at a syntactical level; the students have no 'meta-control'; they know that they are allowed to do some things and not others, and obviously they sometimes make mistakes ... to improve the situation one can call to mind an algebra which is always linked to a context; not necessarily to the (often unreal) 'real world problems', but to the properties of numbers, or to the manipulation of functions, in all cases where it is necessary to interpret the result. (Menghini, 1994, p.13)

We see the important task as making symbol representation meaningful rather than, as a submission to the Cockcroft report (1982) expressed, that:

> Mathematics lessons are very often not about anything. You collect like terms, or learn the laws of indices, with no perception of why anyone needs to do such things (para. 462).

> ...it was the lack of this (linking symbols to the situations they represent) that led to failures in the past teaching of algebra: the children who failed thought of x and y as meaningless marks that had to be played with by peculiar rules (Sawyer, quoted in Anderson, 1978, p. 20).

One argument is that this meaning might be achieved through working with students on thinking mathematically, where algebra is one component:

> One major part of the effort to reform secondary school mathematics is the project of changing the goal of studying school algebra from mastery of symbolic manipulations to the ability to reason mathematically (Yerushalmy, 1997, p. 431).

Students need some fluency in symbolic manipulation, however:

> The manipulation of symbols is only a small part of what algebra is really about, the traces that are left behind after mathematical thinking has taken place. (Mason, 1992, p.5).

One implication of this is that:

> symbolic manipulation should be taught in rich contexts which provide opportunities to learn when and how to use those manipulations. (Arcavi, 1994, p. 32)

In other words, algebra should arise from complex situations:

> Algebraic symbolism should be introduced from the very beginning in situations in which students can appreciate how empowering symbols can be in expressing generalities and justifications of arithmetical phenomena ... in tasks of this nature, manipulations are at the service of structure and meanings. (Arcavi, 1994, p. 33)

There is never an end-point in this conception of learning mathematics. If I am learning to reason mathematically, to structure my thinking about problems, then what I learn is in an ongoing state of complexification and enrichment.

Here we had our link to the challenge (Sutherland, 1991) of creating a school algebra culture in which students find a need for algebraic symbolism. The need we envisage here is for expression of awarenesses within complex situations. This clearly places onus on us as teachers to create a classroom culture in which there is the possibility for students to work at and attempt to express what they are aware of. What we are prepared to notice and able to perceive is to a large extent dependent on the culture around us, and the language available to us.

A unifying strand through all these quotes is the sense of algebra as an evolving language that can emerge from situations and contexts that are already laden with meaning. Algebra can be used to express and offer insights into those situations. It is in this emergent expression and consequent empowerment that students can discover a need for algebra.

We have taken the following definition of algebraic activity from the Sutherland report:

> (i) Generational activities which involve: generalising from arithmetic, generalising from patterns and sequences, generating symbolic expressions and equations which represent quantitative situations, generating expressions of the rules governing numerical relationships.

> (ii) Transformational activities which involve: manipulating and simplifying algebraic expressions to include collecting like terms, factorising, working with inverse operations, solving equations and inequalities with an emphasis on the notion of equations as independent 'objects' which could themselves be manipulated, working with the unknown, shifting between different representations of, function, including tabular, graphical and symbolic.

> (iii) Global, meta-level activities which involve: awareness of mathematical structure, awareness of constraints of the problem situation, anticipation and working backwards, problem-solving, explaining and justifying. (Kieran, quoted in Sutherland, 1997, p. 12)

Within the discussions of the working group set up to write the this definition was the one which covered every member's interpretation of what algebra is. Such a broad definition allows us, as teachers, to work on our recognition of what algebra is in what the students do.

Through trying to identify aspects of the third of these components both in our practice of mathematics and in that of the students, we have come to see algebraic activity as synonymous with thinking and working mathematically. We see our (Alf as teacher, Laurinda as researcher) task in the classroom as developing a community of practice (Lave and Wenger, 1991) where the practice is not that of mathematician, since there may only be one of these (the teacher), but of inquirers (Schoenfeld, 1996) into mathematics. To do this we had to consider how the culture of the classroom, where algebraic activity is present all the time, would be established from the start of the academic year 1998/9, as the students (Alf's new year 7 class) began their life in a secondary school.

An account of these first lessons was presented at a BSRLM meeting in Leeds (Coles and Brown, 1998) where we described the use of the purpose (Brown, 1997; Brown and Coles, 1996; 1997) articulated to the students of 'becoming a mathematician'. The students contributed their own meanings to 'becoming a mathematician' both implicitly (Claxton, 1996) and explicitly through writing about 'what I have learnt'. Alf's role was both to act as a role model of inquirer and to metacomment on the 'mathematical' behaviours of the students as they worked. In the rest of this paper we explore in more detail the meanings we give to the idea of metacommenting and present evidence for such metacommenting by the teacher in the development of a community of inquiry which is supporting the students in finding a 'need' for algebraic activity.

METACOMMENTING

Bateson (1972) says that human verbal communication can operate and always does operate at many contrasting levels of abstraction (p. 177). One of these levels he calls *metacommunicative* (p. 178) when the subject of discourse is the relationship between the speakers. We were interested in what the students did in relation to the task of 'becoming a mathematician' and needed to talk explicitly with the students about our relationship to the practice of inquiry into mathematics. To illustrate metacommunication, Bateson gave the following story:

> What I encountered at the zoo was a phenomenon well known to everybody: I saw two young monkeys playing, i.e. engaged in an interactive sequence of which the unit actions or signals were similar to but not the same as combat. It was evident, even to the human observer, that the sequence as a whole was not combat, and evident to the human observer that to the participant monkeys this was "not combat". Now, this phenomenon, play, could only occur if the participant organisms were capable of some degree of metacommunication, i.e. of exchanging signals which would carry the message "this is play." (p. 179)

Obviously these messages are non-verbal and implicit and in fact Bateson comments that the vast majority of "... metacommunicative messages remain implicit" (p. 178). In the early days of establishing the culture in the classroom, however, we would need the messages to be explicit when the teacher comments to the students about their mathematical behaviours, because, although we believe the students are all naturally inquirers, at this stage they do not know the culture of doing mathematics which they are entering. We have come to call explicit metacommunicative messages metacomments.

HOW WE WORK

Our methodology is situated within and draws upon what Bruner (1990) called a culturally sensitive psychology:

> (which) is and must be based not only upon what people actually do but what they say they do and what they say caused them to do what they did. It is also concerned with what people say others did and why ... how curious that there are so few studies that (ask): how does what one does reveal what one thinks and believes. (pp. 16-17)

In collecting data about the developing culture we concentrated on what Alf and the students did:

• From the start of the year Alf metacommented to the whole class describing his interpretation of what he saw students do in relation to the practice of becoming a mathematician, e.g.:

> This group had an idea which they wrote down and tested and found it didn't work, so they changed their idea. That's a great example of what it is to think mathematically. (Research diary, September, 1998)

Once a week Laurinda would observe the class and her focus of observation was metacomments.

• At the end of a sequence of lessons the students reflected and wrote on "What have I learnt?" which could be about content and also about 'becoming a mathematician'. Sometimes the students would write about something which Alf had explicitly stressed such as "I've learnt to look for patterns" and in other cases they developed their own statements "It's OK to make mistakes." This sense of the students developing their own, several, meanings in relation to 'becoming a mathematician' and Alf not forcing a particular culture, simply staying with describing what was happening for himself or the students, was important to us since our focus was on the students' own need to use algebra.

The research lasted for one term, September to December 1998. The data set of the project is: interviews with 6 students (selected across the range of attainment) at the beginning and end of term, students' exercise book, writing on "what I have learnt?", a half-term review, base-line entry test data, notes from Laurinda's observations of class lessons and Alf's research diary containing notes on planning and reflections.

We are interested in focusing on what we and the students in our classroom do and we use an enactivist methodology (Reid, 1996, Brown and Coles, 1997, Hannula, 1998) which has two key features. The first is that we take multiple views of a wide range of data:

The aim here is not to come to some sort of 'average' interpretation that somehow captures the common essence of disparate situations, but rather to see the sense in a range of occurrences, and the sphere of possibilities involved. (Reid, 1996, p.207)

In practice this means that after every observation visit by Laurinda we talk through the events of the lesson, initially staying with the detail of our observations (accounts of) and moving to interpretations (accounts for). We do not aim to arrive at common interpretations but rather to extend our range of possible actions in the classroom ('noticing paradigm', Mason, 1994). The second key feature of enactivist methodology is:

the importance of working from and with multiple perspectives, and the creation of models and theories which are good-enough for, not definitively of. (Reid, 1996, p. 207)

theories and models ... are not models of. That is to say they do not purport to be representations of an existing reality. Rather they are theories for; they have a purpose, clarifying our understanding of the learning of mathematics for example, and it is their usefulness in terms of that purpose which determines their value. (Reid, 1996, p. 208)

We see our "research about learning as a form of learning" (Reid, 1996, p.208) where our learning is gaining a more and more complex set of awarenesses about our teaching.

The purpose of these theories is to transform our views in the act of staying with the detail of what the students do.

EVIDENCE OF DEVELOPING ALGEBRAIC ACTIVITY

We did not know, when we started, how the culture would develop in relation to Alf metacommenting on behaviours, but what follows is evidence for the developing algebraic activity of the students through this process. We are now starting to classify such behaviours and, in turn, are recognising other instances of them:

Strand 1 – students displaying behaviours implicitly and autonomously which have previously been metacommented on by the teacher.

Strand 2 – students displaying behaviours implicitly and autonomously which have been metacommented on by students in their writing.

Strand 3 – students explicitly metacommenting within the culture of the classroom.

The episodes that we use below are chosen because we have seen similar evidence in other lessons and they seem to us to stand for descriptions of algebraic activity of the students which are now part of the culture. This does not mean that each and every person in the class would exhibit every such behaviour but that we are "coemerging" (Reid, 1996, p. 203) with these behaviours as part of the culture.

We are aware that the majority of the behaviours above will become implicit but we are interested in investigating whether it is possible to establish a community of inquirers in this way, through which the learning of mathematics is facilitated in a holistic manner, underpinned by algebra.

Episode One – evidence of strands 1 and 3

The following interchange occurred during a lesson when the class was looking at drawing graphs of 'rules' such as times by three. A confusion had arisen about the difference between squaring and multiplying by two and between cubing and multiplying by three. The notation of powers for squaring and cubing had been discussed briefly in the previous lesson.

Claire: If you had N, N four, then you just have to put N times N times N times N.

AC: Exactly but that would be different from timesing it by four, those are different rules.

Claire: Because you can, if you write it out like that, you can have, there's a pattern because it's got two and then it's got three then it's got four and it keeps going up like a pattern (Transcribed from videotape, December 1998).

'Noticing pattern' is a behaviour Alf had metacommented on from the first lesson, as being a part of what it is to think mathematically. In this incident Claire is both in the action of extending the pattern (Strand 1) and able to comment on it (Strand 3). There is also evidence of two of Kieran's (quoted in Sutherland, 1997) components of algebraic activity. The sense of spotting pattern and being able to continue a pattern is generational, however what Claire is actually talking about is form and equivalence, which places her activity as transformational. In a lesson in January there was a strikingly similar incident in which Claire built a sequence from the link between $3N$ and N add N add N, which led to $aN = N + N + ... + N$ (a times). On the baseline tests at the start of the year she came out as one of the weakest in the group and is on the school register of students with special educational needs, however there seems to be evidence that 'noticing pattern' is now a part of her, a part of what she sees mathematics lessons being about and she spontaneously generated the higher powers of N, seen in this culture for the first time.

Episode Two – evidence of strands 1 and 2

In working on the *Handshakes* problem the class came up against the problem of how can you find a quick way of working out $1 + 2 + 3 ... + 14$, the number of handshakes for 15 people. A student had offered the image:

$$14 + 13 + 12 + 11 + 10 + 9 + 8$$
$$1 + 2 + 3 + 4 + 5 + 6 + 7$$

Part of the conversation was as follows:

Pupil: 15 goes to 15 times 14 divided by two.

Pupil: This is double.

Pupil: Same as fifteen times seven.

29

Pupil: Equals one hundred and five.

Alex: It all equals fifteen so 'cos there's fourteen numbers you times and then halve.

Ben: That's the same as what Paul said and then halve it.

Shelley: I've got something that I think will help. Can I come to the board? ... You could do it like this: fifteen add fifteen add fifteen add fifteen add fifteen add fifteen add fifteen, but times is quicker, it's fifteen times seven.

Both Alf and Laurinda were confused during the start of this discussion as were other students. The culture, however, supports meaning-making for all participants and since there was a lot of energy to try and sort this out, Alf let the discussion run on. The students were genuinely inquiring into this problem and offering their different ways of seeing it. Shelley's comment shows that she sees supporting others as being part of the culture of the classroom and this is in fact something she has explicitly written about before (Strand 2):

> I think we have been working as mathematicians as a class because we have been sharing our problems and solving each others problems.

Here she is implicitly offering a structural insight which is evidence of Strand 1 and is working at a global meta-level algebraically as she works on how to convince the others in the class.

Episode Three – evidence of strand 3

After we had worked for some time on the problem in episode two Ben asked:

Ben: Is there any way you could do N in this?

This comment took the discussion forward and is Strand 3 because Ben sees that there might be the possibility of going to the general structure and is able to comment on this explicitly.

There seems to be a direct link here between Strand 3 behaviour and the need for algebra. It is through being explicit about a behaviour that is part of the practice of the group ("doing N for this") that a need for algebra is created.

ANALYSIS OF ALL THREE EPISODES

Within our enactivist methodology we are not concerned with establishing causal relationships, but what we observed over the period of one term was Alf metacommenting on 'mathematical' behaviours, these behaviours increasing in frequency in our observations of students and, towards the end of the term, some students metacommenting themselves on mathematical behaviour, often in the form of questions. It is through these Strand 3 behaviours and questions that we have evidence of students needing algebra. In episode 3, Ben created a need for algebra from his question, which, in this instance, Alf responded to by working with the whole group on a general solution.

In the final section of this paper, we illustrate further the connection, in evidence in Episode 3, between a 'need' for algebra and strand 3 behaviour (in particular, asking questions), through a case study of the developing algebraic awareness and mathematical insights and competence of one of the members of this class, Alex.

THE CASE OF ALEX: NEEDING TO USE ALGEBRA

This case study illustrates one student's developing use of the three components of algebraic thinking (Kieran, in Sutherland, 1997), specifically leading to one student (Alex) finding their own need for algebra. Alex was one of the 6 students selected to be interviewed, and this case study takes evidence from both interviews (at the start and end of term). Alex was chosen for this study because he showed the clearest evidence for needing algebra of the students interviewed. He is a high attaining student relative to the group and was one of only five students to score above the national average on standardised baseline tests at the start of the year.

The first activity the class tackled was a rich numerical problem that lasted for seven lessons. In that time Alf was using strategies to allow students to raise many questions within the group although everyone also had a lot of practice with the processes of basic addition and subtraction.

Stage 1: Algebra introduced by teacher but not used by student

Some of the questions students raised involved wanting to know why 9s fell in particular places in the calculations. Alf recognised that one way of answering these questions was to use an algebraic demonstration, since no student was using algebra. Alex had not used algebra before. On being interviewed after the seven lessons he remarked that "basically all of it in my primary school was sums" and further that ideas of proof were not used at primary school. After the demonstration, 11 out of the 27 students could recreate the manipulations and 8 were able to extend the techniques to show other similar results within the problem. We did not, however, expect students to be able to reach for algebraic technique in a different context (the situatedness of learning, Lave and Wenger, 1991) nor were we concerned that the majority of students might not be able to reproduce the original demonstration at this time. The possibility of using algebra to know why things work as they do was now within the community of practice (Lave and Wenger, 1991) and, from another viewpoint, the zone of proximal development (Vygotsky, 1978) of the students, and this was our main purpose in introducing the algebra. At this stage Alex thought of thinking like a mathematician as "you've just got to ask yourself why is it doing this?"

In the first interview Alex was invited (by Alf the interviewer), to try numbers in a problem which he had not seen before. He quickly spotted a difference of 3 (see figure 1).

When Alf asked: "You said that being a mathematician is about asking questions so what's your immediate question?" Alex replied: "Does that happen with every number you put in?" This comment is Strand 3 since Alex is commenting on the process of working on the problem. From an outside perspective we can see that Alex needs algebra to answer his question, however, in working on this question he tried out 'minus numbers' and decimals. It was evidently not natural to let a letter stand for any number and explore the consequences, Alex himself did not recognise he had a need for algebra.

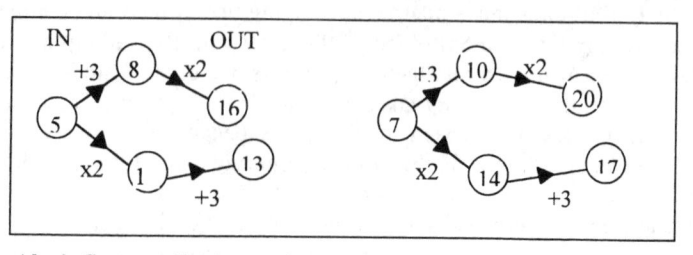

Figure 1: Alex's first two 'both ways'

There is evidence of generational activity here since the pattern of 'there's always a difference of three' was spotted. But, despite recognising 'why' as being a mathematical question, Alex does not ask himself 'why' in this context and consequently does not display global meta-level awareness within this problem. Algebraic symbolism was not used so there was no evidence of transformational activity either.

Stage 2: Algebra used by student in response to teacher's question

Figure 2 is taken from a half term review given to the class which involved some questions to explore how they were getting on with algebra and an end of half-term "what have I learnt?". Here, in response to the prompting in the text of the question, Alex is able to work through the problem using a general letter N (even though there is no explicit invitation to use N in the statement of the task) demonstrating some transformational skills.

We believe he is able to share $2N + 4$ by 2 to get $N + 2$ because of awarenesses formed through the numerical process of trying a few examples first. This is Strand 1. Alf has commented on and exemplified the use and power of algebra in proof and here Alex is displaying such behaviour. Alex is effectively using the skill of multiplying out brackets, but no algorithm for this has yet been taught. Alex recognises that the sequence of instructions always results in 2 and so uses generational activity.

In commenting: "most are alimaneting its selve" (we think this means "eliminating themselves") we would interpret a global meta-level appreciation of the structure of the trick and in reaching for the N also a global meta-level awareness of the power of using symbols, although this happens in response to another's questioning.

3) Try out this trick with different numbers ... write down anything you notice ... can you prove anything about this trick?

Think of a number	*1*	*5*	*100*	*N*	*The Answer always comes out as two.*
↓					
Double it	*2*	*10*	*200*	*2N*	*If you look at the sequence most are aliminating its selve eg Think of a numer, Take away the numer*
↓					
Add 4	*6*	*14*	*204*	*2N+4*	*you first thought of !!!*
↓					
Halve your answer (share by 2)	*3*	*7*	*102*	*N+2*	*same with Double it and Halve your answer, if there was no "add 4" it would come to 0*
↓					
Take away the number you first thought of	*2*	*2*	*2*	*2*	*but there is a "add 4" when that is halve it leaves you with "2"*
↓					
ANSWER	*2*	*2*	*2*	*2*	*the answer.*

Figure 2: Question 3 from Alex's half term review

Stage 3: Algebra needed to answer a question posed by student

In the second interview with Alex, at the end of the first term, Alf posed the same problem as in the first interview, but with different numbers. Alex tried one more example and commented: "The one I've just done was 6 difference and the same for that one there." As in the first two incidents, Alex displays generational activity in noticing a numerical pattern. In response to: "What questions are around for you as you notice a pattern like that?" He replies in a similar way to before: "Does it work for all of them?" This, again, is a Strand 3 comment. Previously this statement led him to try out decimals and 'minus numbers' but after one more numerical example, without speaking, this time he produced the following algebraic solution (figure 3).

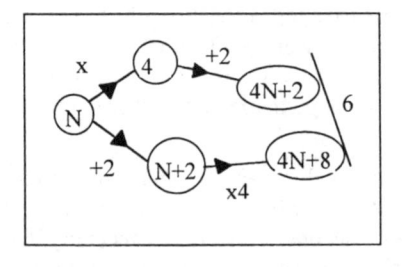

Fig 3: Alex's algebraic 'both ways'

There is certainly evidence here of transformational activity because Alex gains control of the process before using algebraic skills. Even more surprisingly, Alex returned to a numerical problem and said:

> I know what's making it 6 difference now, with the N. Because the bottom way – I can't say it. But that 7 it's going to be more than just timesing it by 4 straight away and adding two on the end. Really you're timesing the 2 plus the 5 by the 4 that way. It's hard to explain. So, that one would be on plus 8. So, these two cancel out each other leaving 6 behind. So now you know every one's going to go to 6.

Alex clearly shows evidence here of insight into the structure of the problem, a global meta-level awareness, which, unlike at the half-term review, is also articulated. As with Stage 1, in recognising a pattern and asking himself "why?" in this new context he creates a need. This time he recognises that need himself. His experiences over the term allow him to answer this need with the use of a letter N to stand for a general number. As he worked through the general case the structure of the problem was illuminated: "I know what's making it 6 difference now, with the N."

In commenting on the process of his solution, Alex recognised the power of N standing for any number: "I should have done that first off." Alf, in reply during the interview, tells him that it is good to start with the process and we would argue that Alex's need for algebra came through the posing of his own question: "why?" and that this came out of a pattern spotted (generational activity) after the process of doing a few examples.

His transformational skills, in contrast to the second stage, appear less dependent on numerical awarenesses, since it is in the transforming that he gains structural insight. It is beginning to feel as though Alex will not need to be taught algorithmically many of the transformational skills needed in secondary school, e.g., how to multiply out brackets.

CONCLUSION

Over the course of a term (15 weeks) Alf metacommented on behaviours he noticed in the classroom that were 'mathematical'. As the term progressed we have evidence of these and other mathematical behaviours being exhibited with increasing frequency by the students, in response to working on rich mathematical activities which might develop over two to three weeks' worth of lessons. In encouraging the students to write as they did mathematics (see Fig. 2, for an example) and also reflecting on what they had learnt, both as mathematicians and in the mathematics itself, questions arose.

This paper has focused on the influence we believe the teacher's metacommenting has had on students' use of algebra. It seems significant also that the students' first experience of algebra at secondary school was one of proof in a meaningful context. The proof allowed the class to see algebra do something that could not have been done without it, in that it provided an answer to a question that had come from the students which they could not answer.

We started with the challenge of whether we could develop a school algebra culture in which students found a need for algebra. Our current theory (that is "good enough for", see Reid, 1996, p.207) is that such a need arises when students have their own questions and that such questions are at a meta-level to the students' actual work in solving problems. In order to ask, one needs to recognise that this is an instance when a question is possible and this entails stepping outside the process of simply doing.

ACKNOWLEDGEMENTS

Thank you to the TTA for funding and Kingsfield School for supporting this research.

REFERENCES

Anderson, J.: 1978, *The Mathematics Curriculum – Algebra*. Glasgow: Blackie.

Arcavi, A.: 1994, 'Symbol sense: informal sense making in formal mathematics.' *For the Learning of Mathematics*, *14*(3), 24-35.

Bateson, G.: 1972, *Steps to an Ecology of Mind*. New York: Chandler.

Brown, L. (with Coles, A.): 1997, 'Being true to ourselves, teacher as researcher: researcher as teacher.' In V. Zack, J. Mousley, C. Breen (eds.), *Developing Practice: Teachers' Inquiry and Educational Change* (pp.103-111). Victoria, Australia: Deakin University.

Brown, L. and Coles, A.: 1996, 'The story of silence: Teacher as researcher, researcher as teacher.' In L. Puig, A. Gutiérrez (eds.), *Proceedings of the Twentieth Conference of the International Group for the Psychology of Mathematics Education* Vol. 2 (pp. 145-152). Valencia.

Brown, L. and Coles, A.: 1997, 'The story of Sarah: Seeing the general in the particular?' In E. Pehkonen (ed.) *Proceedings of the Twenty-first Conference of the International Group for the Psychology of Mathematics Education* Vol. 2 (pp. 113-120). Lahti,.

Bruner, J.: 1990, *Acts of Meaning*. London: Harvard University Press.

Claxton, G.: 1996, 'Implicit theories of learning.' In G. Claxton, A. Atkinson, M. Osborn and M. Wallace (eds.), *Liberating the Learner* (pp. 45-56). London: Routledge.

Cockcroft, W.: 1982, *Mathematics Counts*. London: HMSO.

Coles, A. and Brown, L.: 1998, 'Developing algebra - early evidence from a case study looking at the beginning of year 7.' In E. Bills (ed.) *Proceedings of the BSRLM Day Conference at the University of Leeds* (pp. 17- 21). Coventry: University of Warwick.

Hannula, M.: 1998, 'The case of Rita: "Maybe I started to like math more."' In *Proceedings of the Twenty-second Conference of the International Group for the Psychology of Mathematics Education* Vol. 3 (pp. 33-40). Stellenbosch.

Lave, J. and Wenger, E.: 1991, *Situated Learning: Legitimate Peripheral Participation*. CUP: Cambridge.

Mason, J.: 1992, *Supporting Primary Mathematics – Algebra*. Oxford: OUP.

Mason, J.: 1994, 'Researching from the inside in mathematics education: Locating an I-you relationship.' In J. P. da Ponte & J. F. Matos (eds.), *Proceedings of the Eighteenth Conference of the International Group for the Psychology of Mathematics Education* Vol. 1 (pp. 176-191). Lisbon.

Menghini, M.: 1994, 'Form in algebra: reflecting, with Peacock, on upper secondary school teaching.' *For the Learning of Mathematics, 14*(3), 9-13.

Reid, D. A.: 1996, 'Enactivism as a methodology'. In L. Puig, and A. Gutiérrez, (eds.), *Proceedings of the Twentieth Conference of the International Group for the Psychology of Mathematics Education* Vol. 4 (pp. 203-209). Valencia.

Schoenfeld, A. H.: 1996, 'In fostering communities of inquiry, must it matter that the teacher knows "the answer"?' *For The Learning of Mathematics, 16*(3), 11-16.

Sutherland, R.: 1991, 'Some unanswered research questions on the teaching and learning of algebra.' *For the Learning of Mathematics, 11*(3), 40-46.

Sutherland, R.: 1997, *Teaching and Learning Algebra pre-19*. London: RS/JMC.

Vygotsky, L. S.: 1978, *Mind in Society*. London: Harvard University Press.

Winter, J., Brown, L. and Sutherland, R.: 1997, *Curriculum Materials to Support Courses Bridging the Gap Between GCSE and A Level Mathematics*. London: Schools Curriculum and Assessment Authority.

2 THE FREQUENCY OF SELECTION AND RELATIVE EFFECTIVENESS OF DIFFERENT MENTAL CALCULATION METHODS: SOME EVIDENCE FROM THE 1987 APU SURVEYS

Derek Foxman

Institute of Education, University of London

Some results of a reanalysis of the 1987 Assessment of Performance Unit (APU) survey data on methods used by 11-year-olds to obtain answers to mental arithmetic questions are described in this paper. The methods used by the children would almost certainly have been untaught at that time. The categories used in this analysis have been derived from the work of Beishuizen in the Netherlands and in England. The two main mental methods noted are splitting numbers into separate column values and operating on one of the unsplit or complete numbers in the required calculation. A substantial proportion of 11-year-olds who used complete number methods matched the overall performance of 15-year-olds who answered the same questions in the 1987 APU survey of that age group.

INTRODUCTION

It has been a tradition in several continental countries (such as the Netherlands, Germany, Hungary and Switzerland) to focus more on mental arithmetic in early primary years than has been the case in British schools (Beishuizen and Anghileri, 1998; Bierhoff, 1996; Harries and Sutherland, 1998). However, through the 1990s, interest in Britain in children's ability to calculate mentally has been growing (e.g., The Mathematical Association, 1992; 1997), and in the past five or so years has become a focus of numeracy teaching, particularly in primary classrooms (Straker, 1999; Qualifications and Curriculum Authority, 1999). This development raises a number of issues in the approach to teaching the skills required. It is accepted that pupils can themselves invent methods of mental calculation and that teachers should be aware of these and build on them.

Between 1978 and 1987, the mathematics team of the Assessment of Performance Unit (APU), set up by the then Department of Education and Science (DES) in the UK, carried out a series of surveys of 11- and 15-year-old pupils' mathematical performance. The final surveys in 1987 included tests of mental skills. These produced a wealth of data on the mental calculation methods of small, but fairly representative, sub-samples of these two age groups. This was at a time when calculation seemed not to be highlighted in schools, for the results from the full APU samples found that performance in calculation in both age groups declined in the five years after 1982, the year the Cockcroft Report on mathematics education was published (Foxman, Ruddock, McCallum, and Schagen, 1991). It is likely that the mental methods that pupils used in the classrooms in 1987 (that they are unlikely to have been taught), and their relative effectiveness, could provide some useful information for the teaching of mental calculation skills today (note that in reference to different ways of carrying out calculations I use the term 'method', as was mostly done in the APU work, rather than 'strategy' because there is an issue about what constitutes a 'strategy' in mental calculation; see Beishuizen, 1997; Threlfall, 2000).

This paper presents some results of a reanalysis of APU data from the 11-year-olds, using categories derived from the work of Beishuizen in the Netherlands (Beishuizen, 1997) and in England (Beishuizen and Anghileri, 1998). A few comparisons are also made with the limited data available within the APU sample on the mental skills test of 15-year-old secondary pupils.

THE APU SURVEYS

The purpose of the APU was to develop new methods of assessment and to monitor performance in schools in the core curriculum areas. The Unit commissioned independent research agencies to conduct the surveys and report the results. The National Foundation for Educational Research (NFER) was the agency which carried out the mathematics surveys of 11- and 15-year-olds (Years 6 and 11) in the schools of England, Wales and Northern Ireland between 1978 and 1987. Six surveys of each age group took place during this period, annually from 1978 to 1982, and then an interval of five years before the final ones in 1987. With the coming of the National Curriculum and associated assessment, the government of the day decided to terminate the APU programme of surveys.

The APU practical tests

All the APU mathematics surveys included practical tests that were administered orally by experienced teacher-assessors in one-to-one sessions with pupils. The teachers who acted as assessors were nominated by their LEAs and trained to administer the tests by the NFER research team. About 1200 pupils were selected for the practical tests from the main probability sample of about 13000 who took the more standard written tests, and so were reasonably representative of their populations. About 200 schools participated in each practical test sample and assessors spent a day in each school, usually testing six age 11 or five age 15 pupils. Pupils were administered questions from up to three of the dozen or so 'topics' developed for a survey, and one of the topics in the 1987 surveys was a mental skills test. All the topic tests were untimed in order that pupils could be given as much opportunity as possible to demonstrate what they knew and could do; sessions with some pupils took no more than 25 minutes, and with others over an hour.

The mental skills test

The original intention of the practical tests was to assess mathematics carried out with apparatus or materials and many of the tests did have that feature. But the NFER team saw the one-to-one situation, with its oral delivery of questions, as providing opportunities for controlled interaction between assessor and pupil that did not necessarily imply the use of concrete materials. This interaction, especially the 'probe' question, 'How did you get your answer?', provided rich information on pupils' methods. An article by Jones (1975), 'Don't just mark the answer – have a look at the method', was influential in the APU team's adoption of this approach.

An attempt was made in the first survey of 11-year-olds in 1978 to include some calculations which children could choose to carry out mentally or on paper. They were presented with a verbal problem printed on a sheet of paper (e.g., 'Brian has 64p in his pocket. He buys a scrapbook for 28p. How much money has he left?'). Pupils who did the questions in their heads were then presented with the equivalent calculation out of context (e.g., 64 – 28) and asked to do it as they would in class. Different methods were frequently used for the two types of presentation and, sometimes, different answers obtained. Interestingly, some assessors included comments such as "Pupil did not seem to relate the different answers achieved"; "Pupil treated problem and algorithm as two separate entities". These comments suggest that, for some pupils, the numbers and context involved in a calculation is subservient to the method used to do the calculation. When carried out mentally, the questions produced a range of calculation methods, but the recording technique was then insufficiently developed to cope adequately with the data. However, some of the mental calculation responses given to two subtraction questions were described in the first APU report (Foxman , Cresswell, Ward, Badger, Tuson, and Bloomfield, 1980) and they are easily recognisable as belonging to categories used in this present exercise.

A test of mental calculation skills was not attempted again until the final surveys in 1987 when there was a particular incentive to include them: the recommendation of the Cockcroft Report, published five years earlier, that mental calculation, which had been neglected for some years, should return to the classroom. The APU could not report changes in mental calculation performance since 1982 as there had not been a test of those skills in the surveys held in that year. However, there was a clear pattern of changes in the results of the written tests from 1982 to 1987. These showed an improvement in Geometry and Probability/Statistics, but a decline in performance in calculation at both ages in the five years since the publication of the Cockcroft Report. If any schools did introduce mental tests under the influence of the Cockcroft recommendation, they probably took the form of a restricted time test where only the answers to questions were considered by teachers and not the methods used (as is the case with the mental tests introduced into the National Curriculum assessments). It is reasonable to assume that the mental calculation methods used by pupils in the 1987 surveys were untaught, and may well have been unknown to many teachers.

Many of the results of the mental skills tests of both age groups were reported in Foxman *et al.* (1991), but there was little research on mental calculation at the time to which they could be related. Many of the responses to questions were unclassified and it was thought that some of these might now fit into the classificatory schemes developed by researchers in the 1990s (e.g., Beishuizen, 1997; Fuson, 1992;

Thompson and Smith, 1999). The NFER archives yielded the original interview data on the 256 age 11 pupils who took the test, but, unfortunately, have so far failed to uncover the original data from the 353 15-year-old pupils. For the latter there exist the frequency counts of the original much more limited scheme for classifying pupils' methods and the success rates for each question.

The mental skills questions in the 1987 APU surveys and their administration

Some of the questions were the same for both ages (11- and 15-year-olds) and there were others that had certain features in common (e.g., 26 + 7 for 11-year-olds and 26 + 37 for 15-year-olds). The questions for the two age groups are given below listed in the order they were administered, with one exception in the case of the age 15 test: question 11 was in fact the first. This alteration in the order is made so that the same or parallel questions for the two ages could be placed alongside one another. There were two versions of the age 15 test differing only in the form of question 2: in version B the calculation given in version A was in a money context. This enabled a comparison to be made of the responses to the two formats, about half the sample taking each version. The questions are given in Table 1.

	Age 11		Age 15
1	26 + 7	1	26 + 37
2	64 – 27	2	A. 104 - 67 B. I earn £104 and spend £67. How much have I left?.
3	Start at 13 and count on in fours	3	Start at 13 and count on in sevens
4	Start at 28 and count down in threes	4	Start at 28 and count down in threes
5	I buy fish and chips for £1.46. How much change should I get from £5?	5	I buy fish and chips for £1.46. How much change should I get from £5?
6	How much would I pay for 4 tapes costing £1.99 each?	6	How much would I pay for 4 tapes costing £1.99 each?
7	How many 18p stamps can you buy for £1?	7	How many 18p stamps can you buy for £1?
8	I catch a bus at 9.43 am and arrive at my stop at 10.12 am. How long does the journey take?	8	I catch a bus at 9.43 am and arrive at my stop at 10.12 am. How long does the journey take?
9	I took my pulse. I counted 21 beats in 15 seconds. How many beats per minute is this?	9	For this question pupils were presented with a restaurant bill and asked to point out any errors (there were three) and correct them
10	If December the 9th is on a Tuesday, on which day of the week is December 25th?	10	If December the 9th is on a Saturday, on which day of the week is December 25th?
11	16 x 25	11	32 + 33
12	238 + 143		

Table 1: Mental test questions in the 1987 APU surveys (questions in bold are subject to analysis, see below)

The assessor read the questions to the 15-year-olds, but for the younger pupils the questions were printed in a booklet. Pupils were told by the assessor:

I'm going to give you some questions that I want you to work out in your head. The questions are in this booklet. Here is the first question. Please read it to me and then work out the answer. After each one I'm going to ask you how you worked out your answer – that's not because you have got it wrong (or got it right), I'm just interested in how you worked things out.

The booklet used with the 11-year-olds contained one question per page, and the child was asked to read out each question before being asked to do the calculation. The 11-year-olds did not therefore have to memorise the numbers involved in the calculation, but the 15-year-olds did as they were not provided with a visual format – the questions were delivered to them purely orally.

Assessors were provided with a 'script' that included instructions about what to say to the pupil and how to present materials and questions. The response section of a sheet was divided into four parts: the left hand part to record anything the assessor did or said, the right hand section for what the pupil did or said. The upper part was for recording the answer given and the lower part for the response to the probe, 'How did you work out the answer'. Sometimes pupils changed their initial answer when describing their method – working through the calculation again gave them an opportunity to spot a mistake. In response to a probe from the assessor, pupils could elaborate or explain their method further. The initial response and method were recorded as well as any changes in either resulting from a probe, a corrected answer counting as the outcome.

THE REANALYSIS OF THE AGE 11 APU DATA

The response categories used

The responses to the six age 11 questions that are discussed in this paper have been reanalysed using categories devised by Meindert Beishuizen in the Netherlands and in England (Beishuizen, 1997; Beishuizen and Anghileri, 1998). In this way it has been possible to categorise a large number of the responses that were unclassified in 1987. Beishuizen was interested in any evidence that the APU data might provide for the differential effectiveness of the two mental methods used by children in Britain who would have had much less mental calculation training than Dutch children in 1987.

In this article I focus on questions printed in bold in the above lists – those common to the two age groups, and also the questions 2 where the numbers and operations involved have features in common. However, as indicated earlier, the original data for the 15-year-olds has not been found, and the results that exist are the frequency counts of the original much more limited scheme for classifying pupils' methods. I shall, therefore, first describe the responses of the younger pupils and then explore what comparisons may reasonably be made with the available age 15 data.

The three main categories of number manipulation identified (Standard Algorithm, Split Numbers and Complete Numbers) are described below with some initial examples. Only the last two are mental calculation methods, the first being based on paper and pencil algorithms.

Standard Algorithm:

Some pupils using a standard pencil and paper algorithm may well regard the numbers they are manipulating as separate digits, but others may understand that they are dealing with numbers split into particular column values, units and tens etc., as in the following category, Split Numbers. However, dealing with units first is a feature of the addition, subtraction and multiplication pencil and paper algorithms.

The two main mental calculation categories are:

Split Numbers:

The column values (tens and units in the case of pure numbers and the different units in the case of measures) are dealt with separately in this category, with the higher values normally being considered first. Examples of this procedure are:

$64 - 27 = 60 - 20 + 4 - 7$; $26 + 37 = 20 + 30 + 13$;

Change from £5 for a purchase of £1.46 = £5 − £1 + £1 − 46p

The term 'split' refers only to splitting column values and not to other kinds of splitting numbers such as factorizing.

Complete Numbers:

At least one of the numbers in the calculation is left unsplit and operated on complete:

$64 - 27 = 64 - 20 - 7$; $37 + 26 = 37 + 3 + 10 + 10 + 3$;

Change from £5 for a purchase of £1.46 = £1.46 + 4p + 50p + £3

For two-digit addition and subtraction, comparable distinctions between these two main mental methods have been made in the USA by Fuson, Wearne, Hiebert, Murray, Human, Olivier, Carpenter, and Fennema, (1997) and by Thompson and Smith (1999) in Britain. Fuson *et al.* distinguished between 'Separate Tens' (60 − 20 etc.) and 'Sequence Tens' (64 − 20 etc.), while Thompson and Smith described two of the main methods they identified as 'Partitioning' (of numbers into tens and units) and 'Sequencing' (one number left intact and the other split). The mental calculation categories of 'Split Number' and 'Complete Number' in this article includes the distinctions made by these authors and extends them to numbers in measurement contexts and to other operations. The 'Complete Number' category here, however, is broader than Fuson's 'Sequence Tens' and Thompson's 'Sequencing'. For example,

'Complete Numbers' here incorporates 'Compensation' and 'Complementary Addition' as described by Thompson (2000).

Analyses of the responses to questions

In the following analyses of responses to individual questions, various subcategories of the main mental calculation categories, Split Numbers and Complete Numbers, are noted. Work in the Netherlands (e.g., Beishuizen, 1997) suggests that success rates of mental methods may be a function of the way in which the numbers in a given calculation are manipulated. Of the two mental methods, pupils using Complete Numbers are likely to be more successful than those using Split Numbers, and that is a hypothesis explored in the following results. For instance, in subtraction, the 'smaller from larger' bug (Brown and Burton, 1978) is a well-known error when tens and units are split; when applied to the subtraction 64 – 27 the incorrect answer 43 is obtained. In Germany, similar errors are reported for the Split Number method in Lorenz (1993).

Eleven-year-old pupils in Britain in 1987 would have been well-drilled in using standard algorithms and there are plenty of examples in the data of such methods being carried out mentally. Whether they are as successful as true mental methods when visualised mentally, and, if so, in relation to which numbers and operations is also a matter for consideration.

Question 1.   26 + 7

Unsurprisingly, this question proved to be relatively easy for the 256 11-year-old pupils, 90 per cent of them obtaining the correct answer (see Table 2). The success rate was so high overall that there was no significant difference between the main methods used, except the lower success achieved by those pupils using responses that were unclassified. Few pupils saw the need to use the formal algorithm for such a simple addition.

Method	Number of pupils	% sample	% success
Complete Numbers	161	63	93
Split Numbers	65	25	92
Algorithm	12	5	92
Other/Unclassified	18	7	56
TOTAL	**256**	**100%**	**90%**

Table 2: Responses to 26 + 7

Nearly all Complete Number methods involved adding on to 26, the most popular being bridging through 30 (26 + 4 + 3). Over 50 pupils counted on in ones or added five plus the remaining two, and several of these were observed using their fingers to aid the counting. Nine pupils added 10 to 26 and then compensated for 'over jumping' by subtracting three. There were a few examples of adding six onto 27 instead of seven onto 26. One pupil who worked from 27 used a doubling method with compensation: $27 + 7 - 1$.

The most popular Split Number method was the obvious one, $20 + 6 + 7$, but about a dozen favoured a method that appeared to make use of the pupils' knowledge of doubling: $20 + 6 + 6 + 1$. 'Other' responses included ' I just knew it', and those which were so unclear that it was not possible to assign them to the main categories with any confidence.

Question 2.   64 – 27

This more difficult question achieved a success rate of 60 per cent. The methods used were more varied than for question 1. Listed below in Table 3a are descriptions of some of the mental calculation subcategories derived by Beishuizen (1997) and the acronyms he uses to signify them.

Complete Numbers (starting either with 27 or 64)	Acronym	Examples
Adding on to the start number by bridging to 30 and then further adding in tens (counting up)	AOT/A10	$27 + 3 + 10 + 10...$ or $+ 30 + 4$
Adding on to the start number by adding tens or multiple(s) of ten directly (number + 10)	AOT/N10	$27 + 10 + 10 + ..$ or $+30$
Counting down from the start number in tens (number – 10)	N10	$64 – 20$ (or $– 10 – 10) – 7$
Subtracting units first and then tens	U – N10	$64 – 7 – 20$
Counting down from the start number in tens, 'overjumping' and then compensating	N10C	$64 – 30 + 3$
Split Numbers		
Operating on the tens and units separately, but sequentially (ten sequential)	10s	$60 – 20 + 4 – 7$ (valid); $...+7 – 4$ (invalid)
Operating on the tens and units separately and adding the two results (ten-ten).	1010	$60 – 20; 7 – 4$ (invalid)

Table 3a: Classification of responses to 64 – 27

Of the Complete Number methods, direct subtraction (N10 or N10C) was very much more popular than counting up (AOT/N10 or A10), and twice as many of these (50 pupils) used N10 as used N10C (24 pupils). Only 13 pupils utilised AOT/N10 or A10 methods.

Method	Types	Number of pupils	% sample	% success
Complete Numbers	N10, N10C, AOT/N10/A10	83	32	82
Other Complete Numbers	AOT	6	2	16
Split numbers	10s, 1010	69	27	30
Algorithm		74	29	76
Other/Unclassified		24	9	29
TOTAL		**256**	**100%**	**60%**

Table 3b: Success rates of responses to 64 – 27

All the methods listed in the Complete Number row of Table 3b were equally effective, but six pupils used other AOT methods, either by unspecified means or adding on to 27 by one at a time; not surprisingly only one of these six managed to complete the count correctly. Split Number methods had a much lower success rate than Complete Numbers counting in tens because their use frequently induced errors such as taking the smallest digit from the largest in the units column. On the other hand, users of the standard pencil and paper algorithm achieved a high rate of success with these numbers.

The other four questions to be discussed here were all 'in context' dealing with well-known measures.

Question 5.   I buy fish and chips for £1.46. How much change should I get from £5?

All the main methods employed by the 256 pupils for question 2 were identified in their attempts to do this question, but with a different frequency distribution of both type of method and effectiveness. There were two clear types of Complete Number methods used to attempt this question, the counting up methods: AOT/N10 and A10. Beishuizen (personal communication) considers that it is not clear whether pupils using sequential subtraction methods (10s/SUB) were using complete or split numbers as the £1.46 is to be subtracted from a whole number of pounds. Because of this ambiguity sequential subtraction methods have been assigned to a separate row of Table 4. Examples of each type of method in the context of this question were:

AOT/A10	£1.46 + 4p + £3.50, or + 4p + 50p + £3, or + 54p + £3
AOT/N10	£1.46 + £3 + 54p, or £1.46 + £3 + 50p + 4p
Sequential Subtraction	£5 – £1 – 46p, or £5 – £1 – 40p – 6p;
	£5 – 6p – 40p – £1, or 500 – 46 – 100

Split Numbers   two subtractions: £5 – £1 and £1 – 46p;

two countings up: £1→£5 and 46p→£1;

mix of subtraction and counting up (e.g., £5 – £1 and 46p→£1).

Method	Number of pupils	% sample	% success
Complete Numbers	83	32	84
Sequential Subtraction	61	24	61
Split Numbers	65	25	34
Algorithm	26	10	42
Other/Unclassified	21	8	33
TOTAL	**256**	**100%**	**57%**

Table 4: Responses to change from £5 for £1.46

The results showed a clear advantage for Complete Number methods, but especially for AOT/A10 which is the 'Shopkeeper's method' (also called Complementary Addition) and so the most practical. Seventy-six of the 83 pupils used this method, achieving an 84 per cent success rate. The Complete Number methods were more effective than Split Numbers or Algorithm, and this is true whether the 10s methods are combined with the Complete or the Split Number method rows of the table. A few pupils used personal experience, for example: "I normally get £3 change when I buy things costing £1 something for my mum so I just took away 46 from £1".

Question 6:   How much would I pay for 4 tapes costing £1.99?

By far the most popular Complete Number method (86 pupils) was rounding up to £2, multiplying and then compensating for the rounding (see table 5, below). There were several other Complete Number methods used by only a few pupils (13), such as adding sums of £1.99 and doubling £1.99. The variety of Split Number methods included rounding 99p to £1 and dealing separately with the £1 in the given price. One method classified as Split Number rather than Algorithm was one used by 18 pupils who multiplied 99 by 4 instead of the separate digits. Rounding and compensation was one aspect of several unclear methods that have been consigned to the unclassified category along with a range of other mainly incorrect methods (including guessing). Again the Complete Number methods were more effective, although several pupils compensated incorrectly.

Question 7:   How many 18p stamps can you buy for £1?

There is some uncertainty about this analysis relating to the algorithm method. No assessor described a response that went through all the stages of a long division algorithm.

Method	Number of pupils	% sample	% success
Complete Numbers	99	39	78
Split Numbers	81	32	27
Algorithm	25	10	44
Other/Unclassified	51	20	10
TOTAL	**256**	**100%**	**45%**

Table 5: Responses to 4 tapes @ £1.99 each

However, there were two types of response which probably were algorithmic: '18s into 100 (or £1)', and trial multiplication, usually by five, these being the first two steps in the standard long division procedure. These responses have been included in Table 6 as 'Algorithm?'; unfortunately they were not probed by assessors. Complete number methods included counting in twenties (51 pupils, 20 per cent of the sample), or in eighteens (36 pupils, 14 per cent of the sample), or doubling (34 pupils, 13 per cent of the sample). Split numbers included adding tens and eights successively or multiplication by five splitting tens and eights.

Method	Number of pupils	% sample	% success
Complete Numbers	121	47	63
Split Numbers	30	12	57
Unclear: Algorithm?	48	19	71
Other/Unclassified	57	22	11
TOTAL	**256**	**100%**	**52%**

Table 6: Responses to 18p stamps for £1

The most effective methods, apart from those thought to be algorithmic, were Complete Number methods in which the 18p was rounded to 20p. Pupils using these achieved a success rate of 75 per cent. However, it does appear that there was no great difference in the relative effectiveness of the three main categories of response defined in this paper. This could be because the operation involved was division, which pupils find the most difficult of the four operations, and they were less sure about the method options open to them than in some of the other questions. Additionally, the relative success of the split method involving multiplying or counting 5 tens and 5 eights and algorithmic methods in this particular case may well have been due to the multiplier being 5.

Question 8:   I catch a bus at 9.43 am and arrive at my stop at 10.12 am. How long does my journey take?

There were three types of response classified as Complete Number methods, all of them counting on: A10 and N10 versions adapted to time measures, and rounding up to 9.45 or down to 9.40 before counting on (see table 7). The A10 version consisted of bridging to 9.50 and then through 10 o'clock. Some of the N10 versions included 'overjumping' to 10.13 and compensating i.e., N10C. Several of the rounding responses also involved compensation. Four pupils added on to 100 instead of 60 minutes to an hour. Nearly all the Split Number responses seemed confused, and only six pupils attempted a recognisable standard algorithm procedure.

Method	Number of pupils	% sample	% success
Complete Numbers	154	60	70
Split Numbers	71	28	10
Algorithm	6	2	16
Other/Unclassified	25	10	0
TOTAL	**256**	**100%**	**45%**

Table 7: Responses to time for bus journey from 9.43 to 10.12

The results demonstrated that Complete Number methods were selected overwhelmingly for this question and were even more overwhelmingly effective in comparison to other methods. A remarkable number (71 pupils, 28 per cent of the sample) obtained an answer of over one hour; most of them using Split Numbers, algorithm or other procedures in which they operated on the minutes and hours separately.

SOME COMPARISONS OF THE PERFORMANCES OF THE TWO AGE GROUPS SURVEYED: AGES 11 AND 15

As indicated earlier, there were some differences in the presentation of the questions to the older pupils, who were surveyed in November 1987, while the 11-year-olds were tested in May 1987. Whereas each question asked of the pupils aged 11 was displayed before them while they were calculating their answer to it, the 15-year-olds had to hold the figures in memory because the questions were delivered orally. The comparisons below of the currently reanalysed age 11 data with the original summary age 15 data must therefore be considered with caution.

The summary information available for the 15-year-olds consists of success rates and a few frequency counts of methods that were identified in 1987 (see Table 8). These few data can be fitted to the subcategories developed for the current reanalysis of the age 11 data, but not to the main categories of Complete and Split Numbers. However, details of many subcategories are mentioned in the above commentaries on the reanalysed data. There are no data from the older pupils about the success rates for method subcategories.

Question		% success	
* About half the age 15 sample were given the same figures in a money context		Age 11	Age 15
1	26 + 7 (age 11) 26 + 37 (age 15)	90	86
2	64 – 27 (age 11) *104 – 67 (age 15)	60	69
5	Change for £1.46 from £5?	57	84
6	Cost of 4 tapes @ £1.99 each?	45	76
7	How many 18p stamps for £1?	52	75
8	Bus journey from 9.43 to 10.12am	45	76

Table 8: Ages 11 and 15 success rates

The four equivalent questions, numbers 5 to 8 (see table 8), produced a mean difference in success rate between the two ages of 28 percentage points, which is very close to the overall mean difference for the much larger written test samples in the surveys (Foxman *et al.*, 1991). Although the age 15 questions 1 and 2 were more difficult than the parallel ones for the younger pupils, it seems surprising that, especially for question 2, little progress appears to be made in $4^{1}/_{2}$ years of schooling. In fact the actual progress from age 11 to 15 may be rather less than is apparent here because the 1982 primary children, who were in the same cohort that took the age 15 tests in 1987, achieved a mean score on written calculation questions that was 2.4 percentage points higher than their 1987 counterparts. The following is a list of the methods that were coded similarly for the ages:

Question 5:   Change from £5

The two methods coded for the age 15 mental skills test, Sequential Subtraction (counting down), £5 – £1 – 46p (used by 13 per cent of older pupils), and counting up from £1.46 (35 per cent) were selected at about the same level of frequency by the secondary as the primary pupils. Both of these are Complete Number methods and had a high success rate at age 11.

Question 6:   Cost of 4 tapes

The rounding and compensation method (£1.99→£2; × 4; – 4p) was used a great deal more by the older than the younger pupils (74 per cent to 33 per cent).

Question 7:   18p stamps for £1

Trial multiplication, Doubling, Counting in eighteens, and Rounding to 20p were all coded for the secondary pupils. All but the rounding to 20p methods were selected at similar levels by the two age groups. The latter method more than doubled in frequency from 20 per cent to 44 per cent of their respective samples.

Question 8: Bus journey, 9.43 to 10.12

The only comparable method at the two age groups was counting up through 10 o'clock and there was little difference between their frequencies of use.

The data for the 15-year-olds revealed that the successful rounding up methods used for the tapes and stamps questions were selected substantially more frequently by the older pupils. This difference could be due at least partly to teaching, for rounding was a topic on the syllabus for the Year 11 examinations in 1987, but it demonstrates that the older pupils had available more flexible mental calculation methods than their younger counterparts.

CONCLUSIONS

The principal results of the reanalysis of the age 11 data can be seen most easily from the summary table (Table 9), which lists the percentage of the sample of 256 children who selected each of the main categories of method for each of the questions and the percentage success rate achieved for each method. The percentages of the Other/ Unclassified responses have not been included in the table. The question marks under the Algorithm column in the question 7 row ('How many 18p stamps for £1?') are there as a reminder of the uncertainty about classifying responses to this question as algorithmic (see the earlier discussion of the analysis of the responses to question 7).

Question		% sample selecting method			% success rate for method		
		Complete Numbers	Split Numbers	Algorithm	Complete Numbers	Split Numbers	Algorithm
1	26 + 7	63	25	5	93	92	92
2	64 − 27	32	27	29	82	30	76
5	Change from £5	32	25	10	84	34	42
6	Cost of 4 tapes	39	32	10	78	27	44
7	18p stamps for £1	47	12	19?	63	57	71?
8	Bus journey	60	28	2	70	10	16

Table 9: Methods selected and their success rate for the six questions (age 11)

A feature of this table is that Complete Number methods (e.g., N10, A10) were selected more frequently by the 256 11-year-olds than the other main methods for all six questions (in the case of question 5, however, this is still true if the sequential subtraction methods used are regarded as Complete Numbers, but not if they are classified as Split Numbers.) The standard algorithm was the least frequently selected method for four of the questions, and barely more often than Split Number methods for the other two. Furthermore, for these six questions, the hypothesis derived from work in the Netherlands, that mental calculations using Complete Number methods were more likely to be effective than Split Number methods, has been largely confirmed, decisively so for four of the six questions. These were the ones concerned respectively with working out 64 − 27, the change from £5, the cost of tapes at £1.99 each, and the time taken on a short bus journey which passed through a clock hour: three subtractions and a multiplication. A very easy addition, 26 + 7, achieved too high a success rate to separate the efficacy of the two mental methods and, for the stamps question, basically a division, Complete Number methods were only marginally more successful than Split Numbers. In general, however, visualised pencil and paper methods proved more successful than the mental method of Split Numbers, which probably reflects the teaching emphasis in the 1980s.

Children begin early to develop different methods of calculation (e.g., Fuson, 1972; Thompson, 1995), and the results of the present study indicate that a substantial proportion had, by the age of 11, developed effective Complete Number methods, at least for the questions discussed here. The performance of pupils aged 15 was a good deal better, but the final year of compulsory schooling is rather late to attain this level. It is of interest to note in Table 9 that the success rates of the 11-year-olds who used Complete Number methods were very similar to the overall success rates of the older pupils.

Question	% success rates	
	Using Complete Nos. (Age11)	Overall (Age 15)
5	84	84
6	78	76
7	63	75
8	70	76

Table 9: Comparison of the success rates of age 11 pupils using Complete Number methods and the overall success rates of age 15 pupils

The data in Table 9 suggest that an emphasis on such methods in the curriculum might enable future Year 6 pupils to match the mental calculation performance that was demonstrated by Year 11 pupils in 1987. It remains to be seen whether the National Numeracy Strategy (NNS) can achieve this, but, as Thompson (2000) has pointed out, published documents on the NNS (Department for Education and Employment, 1999; Qualifications and Curriculum Authority, 1999) do not stress the important difference between the two main mental calculation methods. This is unfortunate as the demonstration of the effectiveness of sequential counting in tens – Beishuizen's N10 and A10 methods – supports the idea of offering children a sequential model of number such as the empty number line (Beishuizen, 1999) in preference to a place value model like base 10 number blocks which is associated with the less effective Split Number and algorithmic methods.

Other data available for the children in the present study include their written test scores, the time taken to answer the mental skills test, and the assessors' comments on their demeanour and attitude during the test. Many issues for teaching are raised by the current emphasis on mental calculation in schools and there are some further questions that can be examined with the APU data that could also contribute to these issues, for example:

1. What is the influence of context and numbers on the choice of mental method?

2. Are children consistent in their choice of method?

3. To what extent is choice of method related to overall mathematical ability?

These are questions that are being addressed in continuing work on the APU data.

ACKNOWLEDGEMENT

My grateful thanks are due to Meindert Beishuizen of Leiden University, the Netherlands, for his valuable assistance in the classification of the pupil responses to the APU mental skills test and for his support and constructive comments on drafts of this paper (in addition to those of the anonymous reviewers).

REFERENCES

Beishuizen, M.: 1997, 'Development of mathematical strategies and procedures up to 100.' In M. Beishuizen, K. P. E. Gravemeijer and E. C. D. M. van Lieshout (eds.), *The Role of Contexts and Models in the Development of Mathematical Strategies and Procedures* (pp. 127-162). Utrecht: Freudenthal Institute.

Beishuizen, M.: 1999, 'The empty number line as a new model.' in I. Thompson (ed.), *Issues in Teaching Numeracy in Primary Schools* (pp. 157-168). Buckingham, Open University Press.

Beishuizen, M. and Anghileri, J.: 1998, 'Which mental strategies in the early number curriculum? A comparison of British ideas and Dutch views.' *British Educational Research Journal, 24*(5), 519-538.

Bierhoff, H.: 1996, *Laying the Foundation of Numeracy: A comparison of primary school textbooks in Britain, Germany and Switzerland.* London: National Institute of Economic and Social Research.

Brown, J. and Burton, R. R.: 1978, 'Diagnostic models for procedural bugs in basic mathematical skills.' *Cognitive Science, 2*, 155-192.

Department for Education and Employment: 1999, *The National Numeracy Strategy: Framework for teaching mathematics from Reception to Year 6.* London: DfEE.

Foxman, D., Cresswell, M., Ward, M., Badger, M., Tuson, J. and Bloomfield, B.: 1980, *Mathematical Development. Primary Survey Report No. 1.* London: HMSO.

Foxman, D., Ruddock, G., McCallum, I. and Schagen, I.: 1991, *APU Mathematics Monitoring (Phase 2).* London: School Examinations and Assessment Council.

Fuson, K. C.: 1992, 'Research on whole number addition and subtraction', in D. A. Grouws (ed.), *Handbook of Research on Mathematics Teaching and Learning* (pp 243-275). New York, Macmillan.

Fuson, K. C., Wearne, D., Hiebert, J., Murray, H., Human, P., Olivier, A., Carpenter, T. and Fennema, E.: 1997, 'Children's conceptual structures for multidigit numbers and methods of multidigit addition and subtraction.' *Journal for Research in Mathematics Education, 28,* 130-162.

Harries, T., Sutherland, R., with Winter, J. and Dewhurst, H.: 1998, *A Comparison of Primary Mathematics Textbooks From Five Countries With a Particular Focus on the Treatment of Number.* Unpublished report to the Qualifications and Curriculum Authority.

Lorenz, J. H. (ed.): 1993, *Untersuchungen zum Mathematik Unterricht [Research on Mathematics Teaching],* Köln, Germany: Aulis.

Jones, D. A.: 1975, 'Don't just mark the answer – have a look at the method!' *Mathematics in School, 4,* 29-31.

The Mathematical Association.: 1992, 1997, *Mental Methods in Mathematics: A First Resort.* Leicester: The Mathematical Association.

Qualifications and Curriculum Authority: 1999, *National Numeracy Strategy: Teaching Mental Calculation Strategies.* London: QCA

Straker, A.: 1999, 'The National Numeracy Project: 1996-9.' in I. Thompson (ed.), *Issues in teaching numeracy in primary schools* (pp 39-48). Buckingham: Open University Press.

Thompson, I.: 1995, 'The role of counting in the idiosyncratic mental calculation algorithms of young children.' *European Early Childhood Education Research Journal, 3,* 5-16.

Thompson, I. and Smith, F.: 1999, *Mental Calculation Strategies for the Addition and Subtraction of 2-digit Numbers (Report funded by the Nuffield Foundation).* Newcastle upon Tyne: University of Newcastle upon Tyne.

Thompson, I.: 2000, 'Is the National Numeracy Strategy evidence-based?' *Mathematics Teaching, 171,* 23-27

Threlfall, J.: 2000, 'Mental calculation strategies.' In T. Rowland and C. Morgan, (eds.), *Research in Mathematics Education Volume 2* (pp 77-90). London: British Society for Research into Learning Mathematics.

3 ASSESSING EARLY MATHEMATICAL DEVELOPMENT

Ray Godfrey and Carol Aubrey

Canterbury Christ Church University College

The work discussed here is part of an international study involving Dutch, Belgian, German, Greek, Finnish and Slovenian as well as English children aged between 52 and 98 months. The project is co-ordinated by the University of Utrecht and employs the Utrecht Early Mathematical Competence Test. The full international data set is currently being analysed. This paper reports only the analysis of the English data, together with some international data already in the public domain. The data are interrogated to throw light upon (a) the comparison between England and other countries, (b) the relationship between age and mathematical attainment in the English sample, (c) differences between boys and girls and (d) the relationship between attainment in Piagetian tasks and more arithmetical tasks.

INTRODUCTION

International comparison in mathematics education continues to provoke debate, especially around methods of teaching basic arithmetic (see, for instance, Lawler's Politeia Report Comparing Standards, 2000). This debate has had a real impact on practice and is reflected most recently in policy-makers' pre-occupation with a 'back-to-basics' curriculum (see Marks, 1996; Bierhoff, 1996; Robinson, 1997). An example of this is the introduction of the National Numeracy Strategy from September 1999. This is the fourth change to the mathematics curriculum in England within the last ten years, in fact making little change in the content but placing more emphasis on mental calculation and whole-class teaching methods.

Kavkler, Aubrey, Tancig and Magajna (2000), in a recent case study, investigated children's development of numeracy in two contrasting European contexts – England and Slovenia. The study showed that six-year-old English pupils scored significantly higher than Slovene six-year-olds, who were not yet in formal schooling, on a range of arithmetic, mental calculation and problem-solving tasks. However, by seven years Slovene pupils had caught up on some measures and by eight years there appeared to be no benefits from the English early school entry. In relation to this, the limited longitudinal study that we report here, following young children's early numeracy development between five and six years of age in the European context, provided an opportunity to interrogate further the assumption that teaching of numeracy from an early age in England leads to the raising of standards in international terms.

EFFECTIVE EARLY NUMERACY PROVISION

'Effective Provision of Pre-School Education' (see Sylva, Sammons, Melhuish, Siraj-Blatchford and Taggart, 2000), currently the largest study of its kind, is using multilevel modelling to identify and separate the distinct effects of pre-school setting and primary school attended on performance at the end of Key Stage 1 (KS1). At the start of reception and at the end of KS1 children's academic and behavioural competence will be related to those factors known to have a powerful impact on attainment. These include sex, age within the group, socio-economic disadvantage, parental occupation and educational level, fluency in English and ethnic group. These effects and factors can all be distinguished from effects related to schooling.

In fact, since the introduction of standardised assessment of pupil performance at the end of KS1, the search for appropriate and valid ways of measuring schools' performance in the light of pupils' attainments and other indicators has attracted a wide interest in the field. Strand (1997), for example, tracked pupils in Wandsworth schools who completed baseline assessment at the start of reception class in 1992/3 to the end of National Curriculum KS1 assessment in 1995, employing multilevel techniques to assess educational performance of pupils and 'value added' by schools. Girls were found to make more progress than boys whilst pupils on free school meals (FSM) started with lower attainment and fell further behind during KS1. By contrast, pupils with English as an additional language (EAL) caught up with English-speaking peers. Furthermore, examination of school compositional effects showed that on average pupils made more progress in schools with a high proportion of girls and less progress in schools with a high proportion of pupils entitled to FSM, a high proportion of EAL or where the school average on baseline was high.

Sammons, West and Hind (1997) also used multilevel analysis to take account of the impact of pupil background characteristics on attainment at the end of KS1 in a sample of inner-city schools, though this analysis provided a more limited contextualisation of schools' KS1 results because baseline data were not available to control for prior attainment. They found that attainment was influenced by socio-economic disadvantage (measured by eligibility for FSM), sex, lack of fluency in English, and term of birth (especially for the summer-born). For mathematics and science, however, only fluency in English, socio-economic disadvantage and the term of birth had a significant impact.

Perhaps the most significant contribution to research on value-added and pupils' academic progress through KS1 has been made by Tymms and colleagues (Tymms 1996; 1998; 1999; Tymms, Merrill and Henderson 1997; 1998), both from analysis of their own Performance Indicators in Primary Schools (PIPS) data-base and from analysis of data-sets of other local authorities. This work has demonstrated clearly that assessment taking no more than twenty minutes on entry to school can predict later reading and mathematics attainment, with correlations of around 0.7. Their analysis showed that value-added measures are possible for schools, though they concluded that individual pupil prediction remains problematic. They found that large differences between the progress of pupils in different classes were most noticeable in reception classes, but that differences throughout KS1 were also large. They concluded that being a member of a reception class which made a lot of progress had a long-term impact on the pupils concerned. Moreover although sex, ethnicity, eligibility for FSM, age and EAL often made a statistically significant contribution to the multilevel models used, Tymms *et al.* concluded overall that they added nothing to calculation of value-added measures for schools, once baseline scores had been taken into account. Differences between children which might be explained in terms of ethnicity, for example, were equally well explained by differences in baseline scores. This suggests that ethnic differences manifested themselves in the baseline scores. Some of their latest conclusions are that 'baseline assessment explains about fifty per cent of the pupil level variance in Year 2 and a fair proportion of the school level variance' (Tymms, 1999, pp 34-35). Finally, they noted that such assessment continues to evolve – presumably along with our understanding of this area.

These findings raised a number of questions for the study to be reported here:

- what is the relationship between age and mathematical attainment in the English context?
- what are the differences between boys' and girls' attainment initially and in later tests?
- how does performance vary from task to task?
- how does performance of English pupils compare with that of children from other European countries who enter school between six and seven years?

METHOD

Subjects

More than three hundred schools in a large education authority in the south-east of England were informed about the early numeracy project and invited to take part. The response was overwhelming, with only five schools declining the invitation to participate. In the event the researchers attempted to select schools carefully to include socially and economically varied areas, urban and rural areas, large and small schools, with high and low concentrations of children having FSM (used as a crude social index) and special educational needs, as well as a broad band of achievement levels based on national assessments (the results of schools selected ranged from the twentieth to the ninety-fourth percentile). Eventually twenty-one schools took part with groups of ten children (five boys and five girls) from each reception class selected, so far as possible.

	Mean Age (Months)	Standard Deviation	Number of Children
Boys	60.1	3.56	163
Girls	59.8	3.57	156
Total	60.0	3.56	319

Table 1: Sex and Age of Children in the Sample

Where selection was necessary, class teachers nominated balanced groups of able, average and less able children. Information on the ages at the first cycle of testing is shown in Table 1.

57

Materials

Three forms (A, B and C) of the Utrecht Early Mathematical Competence Test (Van Luit, Van de Rijt and Pennings, 1994) were used. Each form comprised eight sub-topics each with five items, providing forty items in total. These topics are listed in Table 2. Forms A and B were designed as parallel and equivalent. Form C is composed of 20 questions from Form A and 20 from Form B.

1. **Concepts of comparison** (between two non-equivalent cardinal, ordinal or measure situations)

2. **Classification** (grouping of objects in a class on the basis of one or more features)

3. **One-to-one correspondence** (counting and pointing to objects at the same time to make a one-to-one relation)

4. **Seriation** (dealing with discrete and ordered entities)

5. **Using number words** (flexibly and in sequence, in this case, 0 to 20, backwards and forwards)

6. **Structured counting** (counting objects in a variety of arrangements)

7. **Resultative counting** (responding to 'how many' questions without the need to point and count)

8. **Applying general knowledge of numbers in real-life situations.**

Table 2: Topics covered in the Utrecht Early Numeracy Test

The topics include both counting tasks and 'Piagetian' logical tasks, such as classification and seriation. Activities such as classifying objects and sorting sets by different criteria, matching identical objects to find equivalent or non-equivalent sets and ordering activities were influenced by Piagetian theory and were presumed to constitute logical operations which underlie more abstract numerical concepts. In recent years, Piagetian tasks have come under increasing attack as cognitive research has delineated children's number skills development in a fine-grained way (see, for instance, Dolman, 1998).

Van den Heuvel-Panhuizen (1996), amongst others, has pointed to doubts that have arisen with respect to the relationship between these skills and the development of arithmetic. In fact, the Dutch realistic approach to counting was first formulated as a reaction to the practice of neglecting counting activities in favour of developing structuralist concepts, that is, logical forms of reasoning such as correspondence, seriation and classification. Dutch mathematics educators consider it important to identify a number of sub-skills in the counting process: verbal control over the number-word sequence, including counting forwards and backwards; resultative counting, which determines amounts and develops separately from control of the number-word sequence; and abbreviated or structured counting of ordered and unordered sets of objects, which may be visible or only partly visible. In the test, these processes are applied to numbers from 0 to 6, from 0 to 12 and then 0 to 20.

The range of activities in the test, including both logical forms and counting activities increases its interest since it allows consideration of the relationship among different sub-topics at specific testing points and over time and for children with varying amounts of formal schooling.

Procedure

Approximately one hundred children took each form of the test, in each testing cycle (first in January/February, then in June/July of English children's reception year and, finally, in January/February of Year 1).

Each child was tested three times using the same Form of the test. As shown in Table 3, there was some sample attrition, though efforts were made to pursue children changing schools and to revisit schools where children had been absent. No significant differences were found at the first cycle of testing between children later lost to the sample and other children.

	Form A	Form B	Form C	Total
First Test	119	100	100	319
Second Test	113	93	93	299
Third Test	107	88	95	290

Table 3: Sample sizes for each Form of UEMCT at each cycle of testing.

Tests were individually administered with each form taking approximately twenty minutes to complete. Most items were orally presented with children responding mainly to pictorial material or, in the case of some of the counting and number tasks, using Unifix blocks. A few items required children to match two objects in a picture using a pencil to link them.

Analysis

The multilevel analysis used for this study incorporated the hierarchical structure of the data collected, with pupils nested within classes, within schools.

RESULTS

This paper presents part of our analysis of the English data. Scores are taken here at face value and no reliability analysis is reported. This will be dealt with in publications on the international data set. With regard to the international data, only information already in the public domain is used.

The scores for the English sample in the first cycle of testing have been analysed by using multilevel least squares linear regression on individual age and sex to look at differences between areas of the county, schools and classes simultaneously with those between individuals. At no stage in the analysis has the variation between socially and economically different parts of the county or between classes been worthy of note. This is to some extent a result of the small number of geographical areas used and the small number of schools that contributed more than a single class.

The effects of the date of starting school were negligible in the context of regression on age. However, a large majority of the children in the sample had started school in the September preceding their fifth birthday and as the testing started in mid-year the sample contained no young summer entrants. No claims are made for the generalisability over late summer entrants of the findings which here relate to younger children entering formal schooling before the term of their fifth birthday. Residual analysis for all models showed no great evidence of departure from normality or contravention of any other model assumptions.

Age

It is, of course, hardly worth considering the scores without taking age into account. Figure 1 incorporates data from individual children in all three cycles of testing and shows an unsurprising pattern in the plot of total scores against age. This might suggest that the differences between scores in the different testing cycles might be simply part of the pattern of age dependency. Different versions of the regression model with combined cycles of testing agreed that scores increased by about 0.92 with every month of age and passed through a mean level of 21.0 at the mean age of 5 years 5 months.

However a rather more effective discrete cycle model (reduction in deviance 45.0, tested as χ_2^2 with p < .0001) suggests a central value of 20.4 and a monthly increase of 0.53 but with scores on the first test 1.4 lower than in the second and scores in the third test 4.4 higher than in the second. The same effectiveness could not be produced by making the effects of age non-linear, for example having the rate of progress with age gradually increasing. In fact in the third cycle progress with age seems to be reduced to 0.41 per month rather than 0.53. Figure 2 shows the regression lines for each cycle of testing superimposed upon the overall regression line.

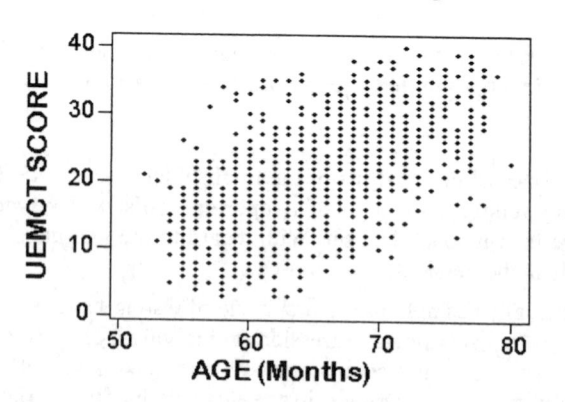

Figure 1: Individual Scores of English Children plotted against Age, superimposing all three cycles of testing.

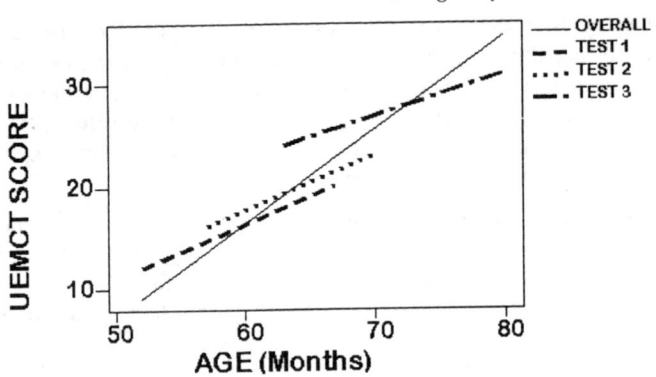

Figure 2: Regression of Individual UEMCT Scores on Age for each Cycle of Testing and Overall

Phenomena similar to this flattening of the age regression lines within each cycle could be artefacts of the method of analysis (see Snijders and Bosker, 1999). However, the difference between the models is so large that some substantive interpretation needs to be made. It is worth noting that a very different pattern emerges from the Slovenian data (Godfrey, Aubrey, Kavkler, Tancig and Magajna, 2000). For Slovene children the same regression line is adequate for both first and second cycles of testing, although there is a difference between these two and the third cycle. The phenomenon cannot be attributed to maturation. It might be due to test familiarity, although children were given no feedback on whether their responses were right or wrong. There is even anecdotal evidence from the testers that many children were becoming more careless with each test.

The effects of the schooling on children of different ages must be considered as a possible cause of the phenomenon. The most obvious interpretation of the present data suggests a different perspective from that usually adopted. While confirming that children who are oldest perform best, the current data suggest that children who are younger than others in the same cohort have their performance enhanced by the presence of older pupils and those who are older have their performance reduced by the presence of younger pupils. At any given time the younger children perform better than the older children did when they were at the age of the younger children. Correspondingly the older children do not perform as well as the younger children eventually will do when they reach the age of the older children.

This suggests that the advantages of the schooling system benefit the younger children in a cohort more than the older children. However, it needs to be remembered that important events in a child's school career happen at certain times of year not at certain ages. When national assessments take place and when children move on to junior classes or junior schools, comparisons are made between children at a particular time although, even under normal circumstances, their ages may differ by up to 12 months. Consequently younger children will be disadvantaged. A further

61

consequence of the current hypothesis would be that age norms should be calculated differently for each time of the year. What counts in January as typical of children then aged 5 years 6 months, is not the same as what is typical in May of children then aged 5 years 6 months. Standardised scores, including those for national tests, that were calculated on the basis of trials early in the school year, may not be appropriate for children being tested in the summer.

The effects of school start date were negligible in this sample. Some studies (for example Sharp, Hutchison and Whetton, 1994; Sharp, 1995; and Sharp and Benefield, 1995) have found this to be an important influence on later achievement. The present result could be due to the LEA policy of admitting almost all children to school in the first term of the school year.

Sex

No appreciable differences were found between boys and girls in terms of the level of scores. When the analysis of the full international data set is published it will indicate that a slight difference in favour of girls is found when some models are applied.

Piagetian v realistic arithmetic tasks

For each of the eight topics, analysis of variance was carried out on the residuals from the regression model, taking all three tests together. That is, allowance was made for progress with age. The analysis looked at the main effects of sex and of cycle of testing. A possible interaction between the two was also investigated, but proved totally insignificant. As with the total scores, even allowing for the progress with age, there was still an impressive difference between cycles of testing in most topics. Although the inspection of the residuals from the ANOVA models apparently suited the assumptions of the model, it is not possible to claim that the data are normally distributed, since there are only six possible scores that a child could receive. These discrete data lead to the standard calculations of significance of the various F statistics giving only approximate answers. Table 4 lists these results. These are different from those presented in an earlier paper (Godfrey and Aubrey, 1999), which used age as a covariate only at the level of groups rather than individual pupils. The comments made on the results shown there are generally supported by the current analysis, though the effects of cycle are relatively reduced here and those of sex enhanced.

Clearly Sex is of very little importance. Even for Correspondence, with the most impressive nominal significance level ($p = 10.6\%$), the effect of Sex is not at all striking. The cycle effects in the separate topics are very significant in most of the counting topics and in Comparison. The other Piagetian topics and Resultative Counting have Cycle effects which do not approach significance, even though when large numbers of hypotheses are tested there is a large risk that some apparently significant effects may be misleading. In this case there is a clear distinction between those effects which are very significant and those which are entirely negligible. In the topics which showed no significant cycle effects, children's progress is approximately in keeping with their age.

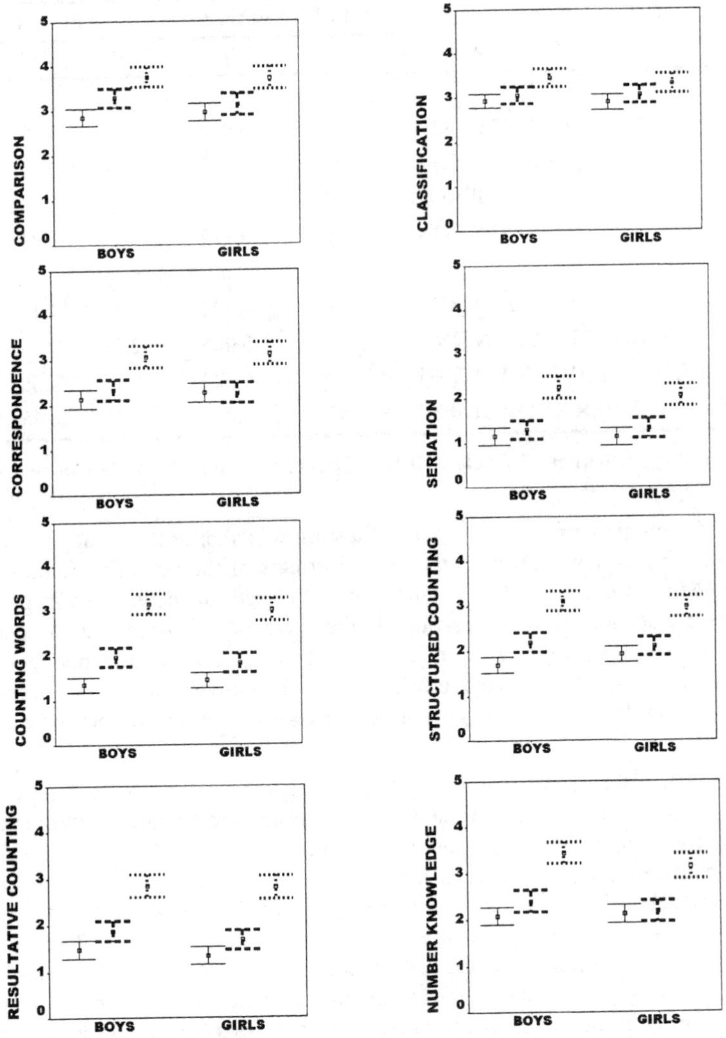

Figure 3: Error Bars showing the 95% Confidence Intervals for the Mean Score in each Topic for Boys and for Girls in each Cycle of Testing.

Figure 3 shows error bars for the mean score of boys and of girls on each of the eight topics in each of the three cycles of testing. It makes clear that some topics contribute more than others to the jumps in performance between cycles. It also indicates the

63

direction of the sex effects given in Table 4 although none of the differences between boys and girls were significant.

TOPIC	CYCLE EFFECT		SEX EFFECT	
	F	"p"	F	"p"
PIAGETIAN				
COMPARISON	6.28	0.002	2.50	0.114
CLASSIFICATION	0.94	0.391	0.09	0.761
CORRESPONDENCE	1.17	0.310	2.62	0.106
SERIATION	1.71	0.179	0.06	0.812
ARITHMETIC				
COUNTING WORDS	11.9	0.000	0.48	0.490
STRUCTURED COUNTING	5.41	0.005	2.07	0.151
RESULTATIVE COUNTING	1.50	0.223	0.19	0.665
GENERAL NUMBER KNOWLEDGE	3.55	0.029	0.17	0.684

Table 4: Nominal Significance of Cycle and Sex Effects in ANOVA Models of Scores for Individual Topics.

The time between the first and second cycles of testing was shorter than that between the second and third cycles. However, the overall effects of the age of the children have already been removed and this means that the effects of differences in gaps between cycles are accounted for. Consequently the cycle effects listed must be seen as jumps between scores on the different cycles of testing. The significance levels shown in Table 4 are all either far above or far below the conventional five per cent level and, for a sample of this size, give a good guide to the importance of each effect.

Figure 3 suggests that the disproportionately large increase in overall scores between cycles two and three occurs mostly in the arithmetic topics and Seriation rather than the other Piagetian topics. The small increases in the Piagetian topics are more in proportion to the time passing between tests. This is consistent with a view that Figure 2 shows schooling producing uniformity of performance. If the school curriculum were directed more towards arithmetic than towards Piagetian development, the effects might show themselves more clearly in the last four graphs of Figure 3. This is in fact what happens. Given that English children go on to perform rather poorly in international comparisons at later ages (see the discussion of Kavkler *et al.*, 2000, above), this could be seen as an unhelpful overemphasis on arithmetic at the expense of other aspects of mathematical development that might ultimately prove more productive.

Figure 4, from Godfrey *et al.* (2000), shows the relationship between progress in the Piagetian Topics and progress in the more realistic arithmetical topics. Comparison of

English and Slovenian children was of particular interest bearing in mind that the English pupils had been in formal schooling throughout and the Slovene children not at all. This shows a far greater bias towards Piagetian topics in Slovenia, where the children were not in formal schooling, than in England, where they were. Age for age the English are rather behind the Slovenians in Piagetian topics, but far ahead of them in realistic topics.

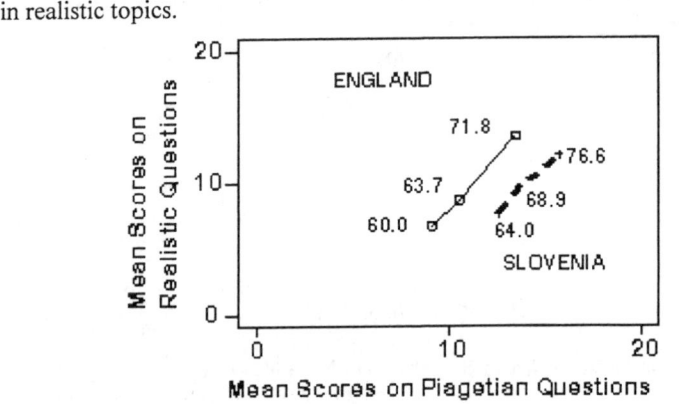

Figure 4: Mean Scores of English and Slovenian Children in each cycle of testing comparing Piagetian and realistic topics, annotated with the Mean Ages.

International comparisons

Van de Rijt and Van Luit (1998) have already reported the results of the first phase of the European study employing the Utrecht Early Mathematical Competence. They presented mean scores for each country in the first test in the cycle. Subsequently Van de Rijt and Van Luit (private correspondence) provided data on national mean ages for the preparation of a symposium paper (Aubrey, Godfrey and Godfrey, 1999). These data are presented in Table 5.

COUNTRY	MEAN SCORE	MEAN AGE (MONTHS)
NETHERLANDS	26.2	71.3
BELGIUM	21.5	66.9
GERMANY	23.5	73.5
GREECE	20.0	67.2
ENGLAND	15.9	60.0
SLOVENIA	20.4	64.0
FINLAND	31.9	80.0

Table 5: National Mean Scores and Mean Ages in the Utrecht Early Mathematical Competence Test.

The ordinary least squares regression calculated at the level of national means shows English pupils performing very much like other nationalities, as indicated in Figure 5. A future publication will report an analysis of the full international data set using appropriate techniques which take into account variability of pupils within countries.

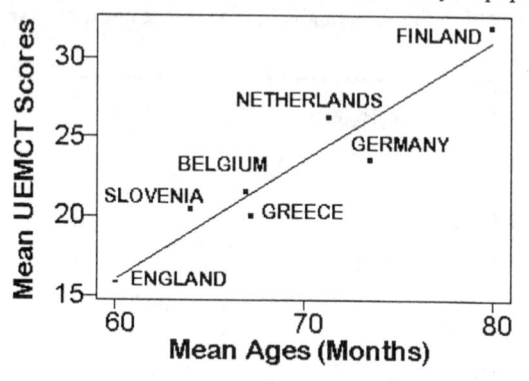

Figure 5: Regression of National Mean Scores on National Mean Age

CONCLUSIONS

The regression models produced for the English data strongly suggest that within a given school cohort, there is a levelling out of age-related differences in performance. The model shows that younger children in a cohort are advantaged rather than disadvantaged, though this does not mean that they will not appear weaker than older children at any one time. It also suggests that age norms in standardized assessments may only be valid for the time during the school year at which they were calculated. No sex differences worthy of note were found.

The English reception classes in this sample appear to produce a larger impact on children's responses to questions on arithmetic than on their responses to Piagetian tasks. Although it is clearly beyond the scope of this analysis to provide evidence of the later consequences of this emphasis, it is consistent with the view that in England the formal curriculum is imposed unnecessarily and unhelpfully early in a child's life.

Although it is possible that some individual questions were more difficult to express in one language rather than another or that the images used for pictorial representations in some questions may have been less familiar in one culture than in another, or that some questions may have presented disproportionate difficulties to the very young children in the English sample, there is no reason to think that English pupils were particularly disadvantaged. The performance of English children was consistent with the pattern in other countries once age is taken into account. Although English children appear at a disadvantage in later comparisons, there is no evidence here of any disadvantage during the first years of formal schooling.

REFERENCES

Aubrey C., Godfrey R. and Godfrey J.: 1999, 'The development of early numeracy in England.' Paper presented at the Conference of the European Association for Research on Learning and Instruction, Gothenburg, August 1999

Bierhoff, H.: 1996, *Laying the Foundation of Numeracy: A Comparison of Primary School Textbooks in Britain, Germany and Switzerland.* London: National Institute for Economic and Social Research.

Dolman, C. (Ed.): 1998, *The Development of Mathematical Skills.* Hove: Psychology Press.

Godfrey R. and Aubrey C.: 1999, 'Assessing early mathematical development.' In L. Bills (Ed.), *Proceedings of the Day Conference held at St Martin's College, Lancaster* (pp. 37-42). Warwick: British Society for Research into Learning Mathematics.

Godfrey R., Aubrey C., Kavkler M., Tancig S. and Magajna, L.: 2000, 'Assessing early mathematical development in England and Slovenia.' *Proceedings of the British Society for Research into Learning Mathematics, 20*(1/2), 85-90.

Kavkler, M., Aubrey, C., Tancig, S., and Magajna, L.: 2000, 'Getting in right from the start? The influence of early school entry on later achievement in mathematics.' *European Early Childhood Education Research Journal, 8*(1), 75-94.

Lawler, S. (Ed): 2000, *Comparing Standards: the Report of the Politeia Education Commission.* London: Politeia

Marks, J.: 1996, *Standards of Arithmetic: How to Correct the Decline.* London: Centre for Policy Studies.

Robinson, P.: 1997, *Literacy, Numeracy and Economics.* London: Wiedenfeld and Nicholson.

Sammons, P., West, A. and Hind, A.: 1997, 'Accounting for variation in pupil attainment at the end of Key Stage 1.' *British Educational Research Journal, 23*(4), 489-512.

Sharp, C.: 1995, 'School entry and the impact of season of birth on attainment.' *Educational Research, 37*(3), 251-265.

Sharp, C. and Benefield, P.: 1995, *Research into Season of Birth and School Achievement. A Selective Annotated Bibliography.* Slough: NFER.

Sharp, C., Hutchison, D. and Whetton, C.: 1994, 'How do season of birth and length of schooling affect children's attainment at Key Stage 1?' *Educational Research, 36*(2), 107-121.

Snijders T. and Bosker R.: 1999, *Multilevel Analysis: An Introduction to Basic and Advanced Multilevel Modelling.* London: Sage.

Strand, S.: 1997, 'Pupil progress during Key Stage 1: A value added analysis of school effects.' *British Educational Research Journal 23*(4), 471-488.

Sylva, K., Sammons, P., Melhuish, E., Siraj-Blatchford, I. and Taggart, B.: 2000, *Technical Paper 1. An Introduction to the EPPE Project.* London: Institute of Education, University of London.

Tymms, P.: 1996, *Baseline Assessment and Value-added. A Report to the School Curriculum and Assessment Authority.* London: SCAA.

Tymms, T.: 1998, 'Starting school: a response to Chris Whetton, Caroline Sharp and Dougal Hutchison.' *Educational Research, 40*(1), 69-71.

Tymms, P.: 1999, 'Baseline assessment, value-added and the prediction of reading.' *Journal of Research in Reading, 22*(1), 27-36.

Tymms, P., Merrell, C. and Henderson, B.: 1997, 'The first year at school: A quantitative investigation of the attainment and progress of pupils.' *Educational Research and Evaluation, 3*(2), 101-118.

Tymms, P., Merrell, C. and Henderson, B.: 1998, *Baseline Assessment and Progress During the First Three Years at School.* Durham: CEM Centre.

Van Luit, J.E.H., Van de Rijt, B.A.M. and Pennings, A.H.: 1994 *Utrechtse Getalbegrip Toets* [Early Numeracy Test]. Doetinchem, Graviant.

Van de Rijt, B.A.M. and Van Luit J.E.H: 1998, 'Development of early numeracy in Europe.' Paper presented at the European Conference on Educational Research, Ljubljana, Slovene, 17-20 September, 1998.

Van den Heuvel-Panhuizen, M. (1996) *Assessment and Realistic Mathematics Education.* Utrecht: CD-B Press.

4 SCHOOL OR COLLEGE MATHEMATICS AND WORKPLACE PRACTICE: AN ACTIVITY THEORY PERSPECTIVE

J.S. Williams, G.D. Wake and N.C. Boreham

Centre for Mathematics Education, University of Manchester

This paper reports on a case study in which we detail how a college mathematics and chemistry student struggles to make sense of the graphical output of an experiment in an industrial chemistry laboratory. The student's attempts to interpret unfamiliar graphical conventions are described and contrasted with those of college mathematics. Our analysis of this draws on activity theory to assist in understanding the position of the student in both the college and the workplace. This highlights the limitations of the experience of the student at college and we question how the mathematics curriculum might be adapted to assist students in making sense of workplace graphical output.

INTRODUCTION AND BACKGROUND

In this paper we report on one case study resulting from a project *Using Mathematics to Understand Workplace Practice* [1] which has been studying the gaps between the mathematical knowledge of students on pre-vocational courses and the knowledge they require to understand workplace practices in vocations to which they aspire. The questions we are interested in include, "What workplace mathematics can we come to understand?", "How is school or college mathematics helpful or unhelpful?" and "What are the obstacles involved, and what strategies might we use to overcome them?" The single case study we report here allows us to illustrate how we have drawn on activity theory to help us understand the difficulties students have in using their mathematics to make sense of workplace practices.

Sometimes the perceived 'deficiency' in students' mathematics as they struggle in workplace situations is characterised as simply a limitation in the amount of mathematics students have studied at school, at other times the quality of their mathematical skills and awareness is questioned. The debate between 'basic skills' and 'authentic performance' protagonists reflects this dichotomy. But our research focuses on the problem in a new way. By studying students on workplace visits and placements as they confront mathematical demands, we examine the social reality in which mathematics and mathematical understanding are embedded.

The catalyst for our analysis has been our involvement in curriculum analysis and design of mathematics courses for non-mathematicians such as those for General National Vocational Qualifications, (GNVQ), see for example Wake and Jervis, (1997), Wake (1997) and Williams, Wake and Jervis (1999). This has involved describing the content and processes which employers and trainers believe to be important. Such research builds on a tradition of syllabus construction which is known to have weaknesses, especially since it relies on perceptions. In the end schools and colleges tend to reconstruct an academic mathematics curriculum much as in the past, thus leaving students with the same curriculum which failed them and an inability to 'transfer' their knowledge between one discipline and another within their study programme and between school or college and the workplace.

A considerable body of cultural and workplace research has sought to understand the problem of how knowledge is related to 'situated activity', i.e. practice (Lave, 1988; Chaiklin and Lave, 1993; Lave and Wenger, 1991; Wertsch, 1991). Ethnographic and cultural studies of mathematical cognition have analysed the mathematics which researchers see in workplace settings, on the street, or 'in the wild', and contrasted this with the type of mathematics which students learn in school (e.g. Nunes, Schliemann and Carraher, 1993; Saxe, 1991). This has led many to innovate in the mathematics curriculum, attempting to develop 'authenticity' in school mathematics practices in a variety of ways (Brown, Collins and Duguid, 1989) though the dynamic of this work tends to a critical approach to school activity as such (Engestrom, 1991; Mellin-Olsen, 1987).

The proponents of the strongest form of 'situated cognition' imply that learning is necessarily so strongly situated that it is folly to teach general academic skills and expect them to be useful in 'real' problem solving. On the other hand some reseachers regard this as ill-founded and overly pessimistic, and provide evidence that knowledge can sometimes be 'transferable', but that transfer requires education and training too, e.g. in general and specific problem solving skills (Anderson, Reder and Simon, 1996). Bereiter (1997) attempts to reconcile these contrary positions through the conception of schools as 'knowledge building' communities, a conception which reflects recent industrial notions of 'learning organisations'. Sfard (1998) also makes a plea for a synthesis of traditional psychological conceptions of mathematical knowledge and the new approach to mathematics as practice.

We aim to shed some light on this problem and ultimately to help develop informed practical solutions to it through this study. The practical problem we are addressing is the need to understand workers' requirements for mathematical competence and to improve our understanding of the links and gaps between what industry requires and what students' mathematical experience has prepared them to understand. This is not just a matter of finding the gaps in students' school mathematics, or ways of improving 'standards' though we do not deny the importance of this! It is essentially a matter of understanding the ways in which mathematics must be understood in the workplace context, and how this differs from the ways in which students in schools and colleges understand and use their mathematics.

The case studies we have been carrying out involve interviews with workers and their managers in an attempt to identify where mathematical demands might be made of visiting students. Similarly we familiarise ourselves with the mathematics the students meet in their school or college courses through field visits and discussions with teachers and students. However, an extra new dimension involved in this research methodology involves introducing the 'voice' of the struggling student (Wertsch, 1991; Engestrom, 1995). This is the vital generative ingredient usually missing in previous workplace research and which we believe has limited the effectiveness of it. Often previous studies have privileged the voice of the worker or the researcher, in identifying the 'mathematics' used in a workplace context. But if bridging the world of school or college to the world of work is in question, what better lens than the student visiting the workplace? Furthermore we also see no other dynamic in previous studies which shows how knowledge gaps might be bridged. The result is, with hindsight, predictable: the gaps seem insurmountable.

Each case study allows us to develop a view of a particular 'world of work' into which we introduce researcher and student to explore how each might best come to understand the worker's activity, drawing on their experience of school or college mathematics. In our analysis, we have used ideas of work process knowledge together with activity theory in which we draw on the 'object of activity' and the mediating 'instruments' and 'division of labour' conceptions of workplace systems in contrasting them with school or college systems. In this paper, we illustrate such analysis with reference to one case study set in an industrial chemistry laboratory; this allows us to highlight some features of the school or college / workplace divide that we have found typical whilst also illustrating some differences in particular mathematical practices relating to graphical interpretation in the two situations.

CASE STUDY: INDUSTRIAL CHEMISTRY LABORATORY

In this case study we investigated how an A level mathematics and chemistry student coped with making sense of chemistry experiments in a workplace laboratory to which she was attached on a week's work-placement. The laboratory carries out a wide range of thermal stability tests, often on a consultancy basis, for manufacturers who have to be sure about safety in their workplace. The student involved was an able mathematician and chemist in the college context, who although only part-way through the first year of her advanced level courses, had in fact studied mathematics beyond this during her pre-16 education overseas prior to her starting her course at a college in the North West of England.

Over the first few days of her work-placement the student had carried out a number of standard thermal stability experiments in the laboratory. These experiments are carried out so that thermal instability is reached as the substance under investigation is heated. During the experiment data is logged real-time and displayed graphically. Discussions involving the student and one of the authors as researcher, and, at a later stage also the workplace supervisor centred round the final printout of graphs of such an experiment, see Figures 1 and 2. In this experiment, a training exercise, toluene (a toxic and flammable liquid often used as a starting material for the manufacture of industrial chemicals) was heated in a flask containing a small amount of metal, and the temperature and pressure were monitored. The student had completed a standard laboratory record of the experiment that made clear that the heating was carried out at a rate of 2 K/min, and that the experiment would cut out if a temperature of 400°C or a pressure of 1000 psig was reached.

Here we identified some of the student's first problems – coping with the unfamiliar units used to measure quantities. Although familiar with the units Kelvin in which the temperature was measured and minutes in which the time was measured, the compound unit (K/min) in which the heating rate was measured was unfamiliar to the student; she would have been more familiar with using degrees celsius (°C) for temperature and seconds for time. The units used to measure pressure – psig, pounds per square inch gauge, which uses standard atmospheric pressure as a zero – were totally unfamiliar to the student who was only familiar with Pascals from her work at college. She was unable to relate the workplace units to those of her own college experience.

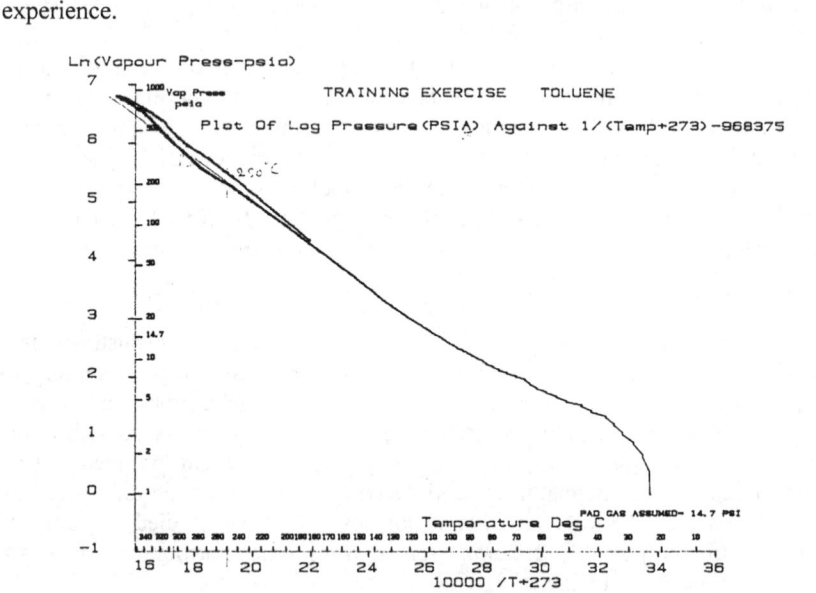

Figure 1: Graphical output of workplace thermal stability experiment, illustrating the use of logarithmic scales and redefined variables.

However the student had even greater difficulty when attempting to interpret the graphs that were the output of her experimental work. Although familiar with graphical work from her studies in both mathematics and chemistry these graphs from the workplace had features for which she was unprepared. Analysis of transcripts of discussions between the student, researcher and workplace supervisor identified these as being:

(i)   more than one variable plotted on each of the horizontal and vertical axes;

(ii)   one of the variables plotted on the horizontal axis having its lowest value at the right-hand end of the axis;

(iii)   the use of a logarithmic scale on the vertical axis;

(iv)   the trace not being an exact straight line.

A particular difficulty the student experienced was due to the fact that in effect the graph in Figure 1 should be 'read' from right to left, i.e. the starting point of the experiment is at the right-hand end of the trace. The industrial chemist creates this difficulty by redefining the variable that is of interest from that of temperature to a coefficient that has importance in analysis of the situation, i.e., 10000/(temperature + 273).

This new variable increases from left to right along the horizontal axis, but this leads to confusion as the new variable is low when temperature is high and is high when temperature is low. Consequently the more familiar quantity, temperature, has high values at the left of the horizontal axis and low values at the right of the horizontal axis. This is the reverse of the student's common experience of graphs in both mathematics and chemistry; so much so that her initial reading of the graph was from left to right and to be consistent with this she was willing to suggest that during the experiment the materials cooled down although she knew that the sample was heated at 2 K/min and hence throughout the temperature was rising. The R(esearcher) and S(tudent) discussed how the graph originated; whether it was traced out real-time as the experiment proceeded or whether it was presented in its finished form at the end. The student confirmed that it was the latter and following further questioning by the researcher attempted to identify the point on the graph at which the experiment started.

R: So where did it start? (The student points to the top left of the trace). It started here, up at this point, 250?

S: Yes.

R: So the temperature got cooler?

S: Yes.

R: Yes? As the experiment proceeded?

S: Yes.

Later in conversation:

S: As the temperature decreases, the pressure also decreases… there will be less pressure.

The vertical axis of the graph proved equally problematic as it used a logarithmic scale. The use of such a scale was unfamiliar to the student although she had a sound understanding of the basics of logarithms (to both base 10 and *e*). It was not clear from the discussion that the student understood that the use of the logarithmic scale

73

effectively hid the rapid increase in pressure during the experiment and the significant instability of the substance under investigation that this represented.

Another aspect of the graphs from the chemistry laboratory that proved problematic to the student was the plotting of multiple traces on the same graph. This particular difficulty became apparent when examining a second graph of the same experiment (see Figure 2). Here the horizontal axis uses the more familiar convention of time increasing from left to right. However, there are now three traces and to make sense of these one needs to use (in this case) two different vertical axes. At the left-hand side of the graph is a temperature scale to which two of the traces (oven and sample temperatures) refer and at the right hand side of the graph is a pressure scale to which the third trace refers. This plotting of multiple variables on the same graph allows one to see clearly how the different variables are affected and interrelated as the experiment proceeds; for example, one can see how:

(i)   the temperatures and pressure all reach a peak at the same time after the start of the experiment;

(ii)   the temperature of the sample lags behind that of the oven.

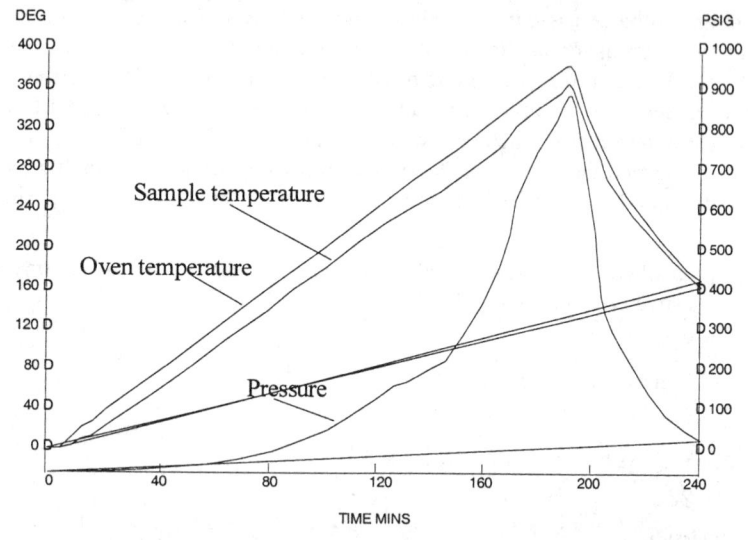

Figure 2: Chemistry laboratory graphs illustrating the use of multiple traces. [2]

The student found interpretation of the graph particularly problematic and the researcher felt that this might be due to the multiple traces plotted on the same graph. He questioned the student as to whether she had ever used similar techniques in her mathematics or chemistry courses in college. It appeared that this was a relatively unfamiliar graphical technique and not one that the student identified as being part of her mathematical experience. The following conversation highlights just some of the student's problems:

R: And what's that outer line?

S: That's the time.

R: Can that be the time?

S: Yes.

R: So why does the time go down?

S: This was the time taken as the experiment was performed, here. But ...

The student in attempting to make sense of multiple traces on one graph was unclear as to why there are two traces of temperature against time, and indeed suggested that one of these is of time itself. She was not uncomfortable with this idea although as the trace is not always linearly increasing this would suggest that after a while time would in fact decrease. The student also found it difficult to locate the starting point of the experiment – she was convinced that in this case the starting point is where all three traces peak. Perhaps this confusion was brought about by having spent some considerable time discussing the graph of Figure 1 and having determined that in this case the start of the experiment was not at the left-hand end of the horizontal axis.

GRAPHS IN SCHOOL OR COLLEGE MATHEMATICS

Let us contrast the use of these graphs in the industrial chemistry laboratory with the graphs that a mathematics student is likely to meet on her school or college course. It should be noted that mathematics courses at this level have in recent years placed more reliance on graphical work and indeed graphs have been used to support teaching and learning to a much greater extent than previously. In particular they have been used to encourage and develop student understanding of a wide range of concepts. This has been due to the advent of relatively cheap computer technology and in particular hand-held graphic calculators. These new technologies have allowed students greater access to use graphs of both data and functions allowing them to concentrate on visualising the mathematical ideas behind graphs rather than having to spend time developing plotting and graph sketching skills. Therefore students currently have in their hands technology that allows them access to graphs as learning tools. This has been reflected by a shift in the nature of the demands of assessment. However, even though there have been these considerable developments, the graphical conventions adopted in mathematics have remained fixed. These conventions that act to assist students in making sense of their school or college mathematics possibly actively add to their difficulties when trying to make sense of workplace situations.

Figures 3 and 4 show graphs (typically found in assessment papers) that a student will meet in mathematics at this level. Notice the 'cleanness' of the situations in contrast to the graphs of the industrial chemistry laboratory.

The graph in Figure 3 is of an undefined function and the work of the student focuses on the formal conventions and mathematical language associated with functions; in this case questions are asked relating to domain, range and inverse. Even when the graph has a context such as that in Figure 4, which shows the acceleration and deceleration of a car over a time interval, the context is not only possibly familiar, but also uncluttered and clean.

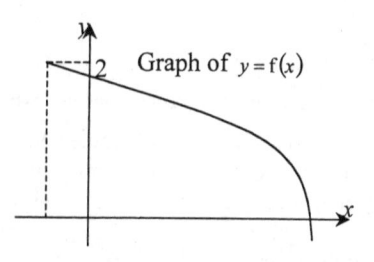

Figure 3: Graph typical of those from the assessment of 'pure' mathematics at A Level.

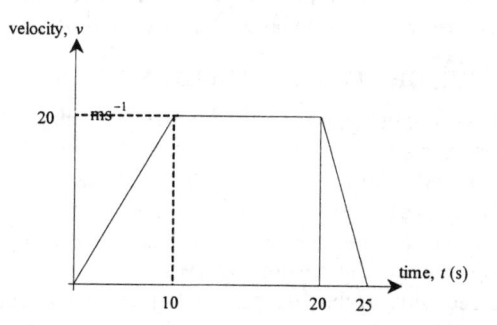

Figure 4: Graph typical of those from the assessment of 'applied' mathematics at A Level.

These figures show graphs that will typically be met by an A level mathematics student. It should be noted that in contrast to the graphs of the laboratory:

- standard labelling conventions are used and where appropriate standard scientific units are used for measured quantities;

- the plotted graphs, straight or curved, follow clear underlying patterns;

- only one variable is represented on each axis.

ANALYSIS: CROSSING THE SCHOOL OR COLLEGE / WORKPLACE DIVIDE

The case study reported here shows that the problems the student faced in using her college mathematics to make sense of graphs arising from the workplace involved an appreciation that the conventions of school and workplace graphs might be different. Indeed the student appeared not to 'see' that college graphs involve somewhat arbitrary conventions and rules.

School or college graphing activities most often involve 'drawing' or 'reading' rather well-behaved and known graphs and functions, rather than 'interpreting' in complex contexts. Thus much emphasis in the curriculum is given to drawing a range of polynomials and other standard functions, and to reading off particular values such as zeros and turning values. On the other hand the graphical output in the chemistry laboratory is read by the expert with a very specific set of conventions which have been moulded to the task of investigating chemical behaviour. These conventions are transparent to the expert but opaque to the novice, because the expert is well placed to understand the context of the experiment through a wealth of past practice.

These problems can be illuminated by reference to studies of 'work process knowledge'. The theory of work process knowledge was developed to explicate ways of knowing in the workplace, and to guide the development of vocational curricula (Boreham, 1999).

It is based on the distinction between science and know-how, the former consisting of abstract, general principles and decontextualised inference procedures, the latter comprising procedural knowledge learnt on the job, especially rules-of-thumb. The fundamental point made by studies of work process knowledge is that the kind of knowledge that guides skilled workers in non-routinised activity is based on a synthesis of these two ways of knowing. This has been demonstrated in several empirical studies of skilled work, including research into prescribing drugs (Boreham, Foster and Mawer, 1992; 2000), production islands in German mechanical engineering companies (Fischer, 1995) and industrial chemistry laboratories (Fischer and Roben, 1997). Evidence was found in all these investigations that the knowledge used by skilled practitioners – their work process knowledge – is constructed in the workplace through a dialectic between know-how learned on-the-job and theoretical knowledge learned in school or college.

The key finding was that work often creates contradictions between these two ways of knowing – for example, the theory learned in school or college may describe an ideal situation, while the real one has idiosyncrasies that do not match the ideal type. The problems experienced by the student in the case study reported here illustrate this contradiction: the graphs taught in college did not possess the idiosyncrasies of reversed time scales, dual plots and so on. Equally, the studies cited reported that "how-tos" learned by experience in the workplace might not be adequate for dealing with novel situations. They found that this contradiction was also resolved by a dialectic between theory and experience, resulting in either a modification of customary procedures, or a revision of the theory, or both. Research in this field indicates that work process knowledge is constructed by resolving contradictions such as these. For example, the study by Fischer (1995) concluded that skilled workers had acquired their work process knowledge by solving production problems, such as diagnosing why the production process was failing with new materials, or deciding how to produce a one-off to a new design. Similarly, in the study of drug prescribing (Boreham et al., 1992), work process knowledge was constructed out of contradictions between procedures learned by experience and theoretical models of pharmacokinetics, which were experienced when individual patients did not respond as expected to a new dose.

This poses the question: why are the conventions of the school or college mathematics course different from the conventions of mathematical activity in the industrial chemical laboratory? This can be tackled by considering the activity system of the college in the sense of Engestrom. Such a system relating to the learning of graphicacy in school or college is schematised in Figure 5.

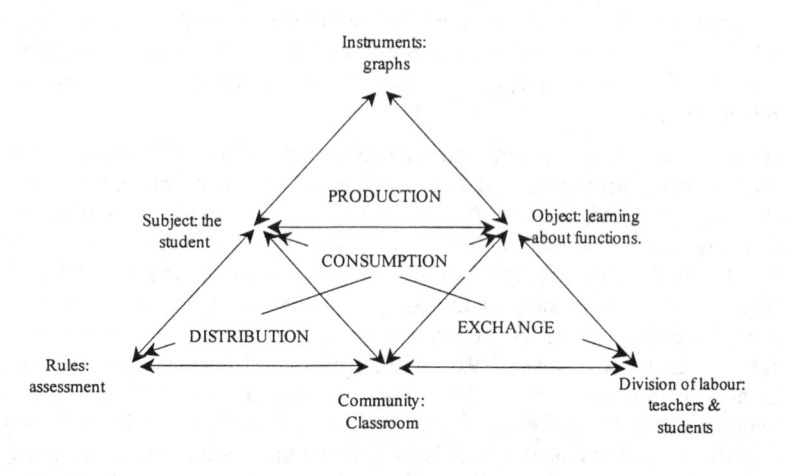

Figure 5: Activity system: the learning of college graphicacy.

Engestrom emphasises that there are key elements within the system which mediate activity: first there are the tools, artifacts and other instruments of action (including conceptual tools). Second there are the rules of performance of the community engaged in a practice (including rules of behaviour and language genres, for instance). Thirdly there is the division of labour in the community and generally the system of the workplace or environment in which the activity occurs.

However, over all these the object of the activity system predominates. This is the goal which the participants are attempting to achieve. A fundamental assumption of activity theory is that their cognitive processes are organised by this objective, and can only be understood by a researcher (or acquired by a learner) who also understands the object of the activity. This is crucial for the present enquiry, because the object depends on the social context in which the activity is taking place. Even if a cognitive tool (such as graphing data) appears the same to an external observer, it will acquire different meanings in different social contexts due to the different objects of the activity, and the different historical processes which have brought them into being.

Consider the activity system in which the student learns and practises mathematics. The purpose of graphs in the mathematics curriculum must be recognised: in part they are intended to provide a tool for studying functions, and familiarity with a simple set of conventions can allow students to 'see' the mathematics behind the graph which unfamiliar conventions would make opaque. Thus when a student moves between the school or college and the work context, and carries the symbolic tool, the graph, with them, they carry also the vestiges of the activity system of school / college, and its historical roots, cultural appendages and other 'irrelevancies'. The student therefore embodies the contradictions of the two systems, as do we all in general when we cross barriers.

The industrial chemistry laboratory operates in a different social context, and so has a different activity system, organised to achieve a different object. In this case, it is to report to individual manufacturers on the safety of substances they have entrusted to the laboratory for testing. Although graphs are used in both college and industry, the meanings they acquire are different in the two settings. This can be explained by examining the complex ways in which symbols have meanings, and especially how these meanings vary with context. This is usually depicted in logic by the distinction between 'type' and 'token'. A graph has both a type-meaning and token-meanings. As a type, its meaning is the same in all contexts – a visual means of presenting relations defined on Cartesian product sets. A token, however, is one physical use of a graph in a particular context. The difference in meaning that might result arises from the fact that tokens may possess indexical meaning; that is, meaning derived from the context of the use. (The common example of an indexical is the indexical meaning of 'I'. The meaning of a particular utterance of this sound – the token 'I' – can only be grasped by finding out who has uttered it.) Now, although the type meaning of graphs, logarithmic scales etc. is the same in college and workplace, the actual use of these terms in the different contexts as tokens these might acquire different indexical meanings. The college activity system is organised around the object of delivering the curriculum and ensuring that the student passes an examination. Thus graphing might acquire a special meaning in that context, such as testing understanding of functions. These meanings might be inconsistent with the indexical meanings acquired by the tokens in the industrial setting. In the latter context, the industrial chemist has redefined the variable of interest from temperature to a coefficient that will enable her to achieve the commercial goal of advising the customer about the safety limits of a new product i.e. 10000/(temperature + 273). The student's confusion arose from the fact that this resulted in a variable that was low when temperature was high and high when temperature was low. Plotting this new variable on the horizontal axis gave rise to a situation that was probably in contradiction with all previous graphs that the student had met. The student transferred her understanding of graphs as a type, but also her understanding of its token use in college which conflicted with its meaning as a token in the laboratory.

The resulting confusion demands a problem solving effort by the student to reconstitute the 'graph' as an effective tool in the workplace. This is in effect learning as re-contextualising (van Oers, 1998), or the process of constructing work process knowledge out of the contradictions between discourses of the college and the workplace (Fischer, 1995).

Figure 6 gives the schema of the activity that situates the student in the workplace where the object is now to make sense of the graphical output of the experiment; these graphs have been problematised by the researcher. We acknowledge that the intervention of the researcher has created an unusual situation in which the probing of the problem of graphical interpretation has pushed the discourse to reflect on the instruments in use (c.f. Leont'ev, 1981). The new rules of the activity include the questioning and probing of the researcher and the student has few available resources with which to make sense of the workplace graphs - the conversations between student and researcher suggest that unsurprisingly she falls back on the conventions and restricted experience of her college mathematics which has not well prepared her to make sense of the use of graphs in non-college situations.

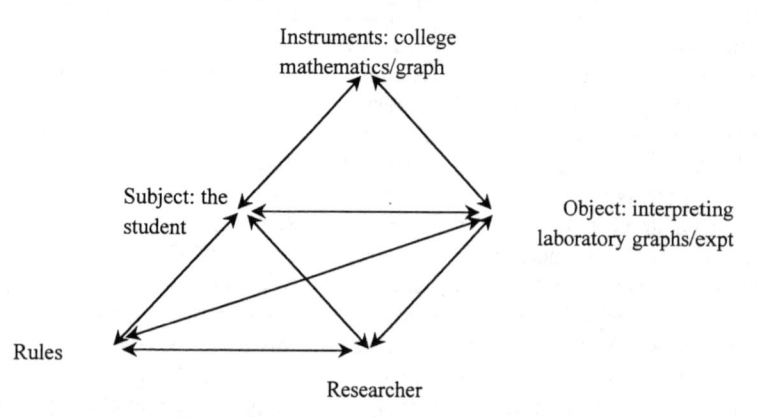

Figure 6: Activity (system): using workplace graphs to interpret experimental results.

It is the student therefore who brings the two activity systems of college and workplace into contact and contradiction and it is consideration of the instruments available to the student in the workplace situation that allows us to draw lessons for the mathematics curriculum. This case study led us to conclude that the conventions upon which our mathematics curriculum is built should be problematised for students. Students would therefore experience working with a variety of conventions and understand that different conventions may develop in different situations. In understanding this, they must develop a sense of the different levels of meaning invoked when mathematical tools operate as types and as tokens. In learning to use mathematics at work, they must become aware of the indexical meanings acquired in those contexts. This involves appreciating that an indexical meaning cannot be derived by logical deduction from more fundamental axioms that define the type. Rather, it can be grasped only by obtaining information from the context in which the token is used, especially about the purposes for which the graph is being constructed, in other words, by understanding the object of the activity system, and how the employees organise their activity to achieve that object. In creating this understanding, the student will be constructing work process knowledge. But it is important to remember that the meaning of a graph in the workplace extends beyond its indexical use. The graph is a type as well as a token; this is the basis of its power to interrogate know-how when contradictions occur between past experience and present problems, an essential part of the process of constructing work process knowledge.

Other case studies we have examined provide further examples of problems encountered by students in making sense of workplace practice due to the use of mathematical tools (types) as different tokens in school or college and workplaces. For example, among others we have observed:

- algebraic notation and expressions,
- coordinate systems,
- percentage errors,
- the use and expression of negative numbers,

being developed in workplaces in ways that do not follow the conventions used and accepted in school or college mathematics classrooms. As we have reported here students' lack of exposure to different tokens of a certain type appear to leave them without resources on which they can draw to make sense of mathematics in new situations.

Finally for us as mediators and observers of this situation, we find the transition from college to the workplace a refreshing exposure of the college 'mathematics' to foreign conditions. One feels like the anthropologists who visit exotic cultures, and with luck learn perhaps a little about their own culture. The identification of contradictions between 'working mathematics' and 'schooling mathematics' provides us mainly with a critique of the latter. The school curriculum rightly emphasises the type meanings of mathematics, which are universal. However, the way the curriculum is implemented seems not to facilitate the ability of students to easily make sense of the indexical meanings which arise when the same mathematics is used in new contexts.

We notice that our interpretation of the data in the case study has led from a social analysis of the activity systems in which mathematics is practised to a critical view of the practice of 'school mathematics' and its values. In this case study and others we have observed that the conventions and rules of mathematical practice in the workplace have developed in ways significantly different from those in which they are taught and practised in school and college. This can be reflected, for example, in artifacts which use different terms, write numbers or graphical results in unusual ways etc. From a school or college perspective these seem 'unconventional'; but this suggests to us that in fact school or college mathematics also has evolved its own idiosyncratic 'conventions' (e.g. about the kinds of functions typically graphed) which are the result of 'schooling' practices and which act to inhibit students' abilities to 'transfer' this mathematics to make sense of new or novel situations. This analysis leads us to conclude that ways should be found to make 'conventional College practices' more explicit, and that the curriculum should be developed to promote the skills required by students to explore mathematical conventions in a range of situations.

NOTES

1. This project was supported by a grant from the Leverhulme Trust to the University of Manchester.

2. The identification of the different traces on the original graph was not so clear. It was achieved using a key referring to the colours of the different traces. It has not been possible to reproduce this here.

REFERENCES

Anderson, J.R., Reder, L.M. and Simon, H.A.:1996, 'Situated learning and education.' *Educational Researcher, 25* (4), 5-11.

Bereiter, C.: 1997, 'Situated cognition and how to overcome it.' In D. Kirschner and J.A. Whitson (eds.). *Situated Cognition: Social, Semiotic and Psychological Perspectives* (pp. 281-300). New Jersey: Lawrence Erlbaum.

Boreham, N.C.: 1999, 'Work process knowledge - the way forward for vocational education and training?' Paper presented to the Annual Conference of the British Educational Research Association, University of Sussex, September 1999.

Boreham, N.C., Foster, R.W. and Mawer, G.E.: 1992, 'Strategies and knowledge in the control of the symptoms of a chronic illness.' *Le Travail Humain, 55*, 15-34.

Boreham, N.C., Foster, R. and Mawer, G.: 2000, "Medical students' errors in pharmacotherapeutics," Medical Education, 34, 188-192.

Brown, J., Collins, A. and Duguid, P.: 1989, 'Situated cognition and the culture of learning.' *Educational Researcher, 18* (1), 32-42.

Chaiklin, S. and Lave, J.: 1993, *Understanding Practice*. Cambridge, UK: Cambridge University Press.

Engestrom, Y.: 1991, 'Non scolae sed vitae discimus: Toward overcoming the encapsulation of school learning.' *Learning and Instruction, 1* (3), 243-259.

Engestrom, Y.: 1995, 'Voice as communicative action.' *Mind, Culture, and Activity.* 2(3), 192-214.

Fischer, M.: 1995, 'Technikverständnis von Facharbeitern im Spannungsfeld von beruflicher Bildung und Arbeitserfahrung. Untersucht anhand einer Erprobung rechnergestützter Arbeitsplanungs- und -steuerungssysteme.' *Schriftenreihe Berufliche Bildung - Wandel von Arbeit und Technik.* Donat Verlag, Bremen. English translation of title: 'How skilled workers understand technical artifacts. The impact of vocational education and work experience.'

Fischer, M and Roeben, P.: 1997, Arbeitsprozesswissen im chemischen Labor. In *Arbeit-Zeitschrift fur Arbeitsforschung, Arbeitsgestaltung und Arbeitspolitik.* Westdeutscher Verlag, pp. 247-266.

Lave, J.: 1988, *Cognition in Practice: Mind, mathematics and Culture in Everyday Life.* Cambridge, UK: Cambridge University Press.

Lave, J. and Wenger, E.: 1991, *Situated Learning: Legitimate Peripheral Participation.* Cambridge, UK: Cambridge University Press.

Leont'ev, A.N.: 1981, 'The problem of activity in soviet psychology.' In J. V. Wertsch, (Translator & ed.) *The Concept of Activity in Soviet Psychology*, p.37-71. Armonk, NY: M. E. Sharpe.

Mellin-Olsen, S.: 1987, *The Politics of Mathematics Education.* Dordrecht, Kluwer Academic Publishers.

Nunes, T., Schliemann, A. D. and Carraher, D. W.: 1993, *Street Mathematics and School Mathematics.* Cambridge, UK: Cambridge University Press.

Saxe, G.B.: 1991, *Culture and Cognitive Development: Studies in Mathematical Understanding.* Hillsdale, NJ: Lawrence Erlbaum.

Sfard, A.: 1998, 'On two metaphors for learning and the dangers of choosing just one.' *Educational Researcher, 27* (2), 4-13.

Van Oers, B.: 1998, 'From context to contextualising.' *Learning and Instruction, 8,*(6), 473-488.

Wake, G.D. and Jervis, A.: 1997, *Mathematics and Science Capabilities of Students with Technology GNVQs.* Cheltenham, U.K.: UCAS.

Wake, G.D.: 1997, 'Improving Mathematics Provision for Post-16 Students.' *Teaching Mathematics and its Applications, 16,* (4), 200-206.

Wertsch, J.V.: 1991, *Voices of the Mind: A Sociocultural Approach to Mediated Action.* London: Harvester.

Williams, J.S., Wake, G.D. and Jervis, A.: 1999, 'General mathematical competence: a new way of describing and assessing the mathematics curriculum.' In C. Hoyles, C. Morgan and G. Woodhouse (eds.). *Rethinking the Mathematics Curriculum, Studies in Mathematics Education Series, 10* (pp. 90-103). London: Falmer Press.

THEME 2
TEACHERS AND TEACHER DEVELOPMENT

The three chapters in this section focus on mathematics teachers, looking at both experienced and new teachers. At a time when initial teacher education in the UK is expected to place increased emphasis on the 'subject knowledge' of trainee teachers, the chapters by Cosette Crisan and by Stephanie Prestage and Pat Perks both address this issue, attempting to go beyond simplistic notions of subject knowledge. Prestage and Perks build a theoretical model of professional knowledge, based on empirical studies of both experienced and pre-service teachers. The importance of 'teacher-knowledge' of mathematics is highlighted, going beyond the sort of knowledge needed to pass examinations to the rather different kind of knowledge needed to help students to acquire knowledge themselves. The authors propose a further elaboration of their model to incorporate an equivalent notion of 'teacher-knowledge' for teacher educators.

The majority of studies of the introduction of computers and other forms of ICT into mathematics have focused on the effects on learners. As Crisan points out, however, teachers are learners too, particularly as they begin to use new technologies in their teaching. In her chapter, she uses a small study of teachers, who have incorporated ICT into their teaching, in order to develop a model to describe their professional knowledge base. This extension of thinking about professional knowledge seeks to account for the ways mathematics teachers use ICT in their teaching and the ways in which using ICT may affect their thinking and practice.

As Jim Smith points out, pre-service courses have generally been found to have limited impact on new teachers' beliefs about the nature of teaching and learning. This might lead to the conclusion that teaching is essentially conservative and that pre-service courses are unlikely to affect the approaches that new teachers adopt to teaching. Through case studies of student teachers' development through their training year, Smith shows that, although their beliefs remained stable, these did not necessarily lead to a restricted approach to teaching. His chapter suggests some ways in which pre-service courses might encourage new teachers to use a wider range of approaches.

5   THE INTERACTION BETWEEN THE USE OF ICT AND MATHEMATICS TEACHERS' PROFESSIONAL KNOWLEDGE BASE FOR TEACHING

Cosette Crisan

Centre for Mathematics Education, South Bank University, London

With the increased availability of information and communications technology (ICT) in schools, it is important to examine how teachers use it in their instruction. This paper describes a study that involved a number of secondary mathematics teachers who had been using the new technology in their teaching for a number of years. It reports on some of the factors affecting teacher' use of the new technology identified by analysing the data collected through questionnaires and teachers' interviews. A possible theoretical framework to account for teachers' learning about ICT and teachers' incorporation of ICT into their planning for teaching mathematics is presented.

BACKGROUND

This paper reports on data from a study of a number of secondary mathematics teachers who regarded themselves as confident and competent users of information and communications technology (ICT) [1] in their mathematics lessons. The purpose of the study was to get a general picture of these mathematics teachers' use of ICT by means of questionnaires and interviews. The analysis of the data focused on some of the factors which were identified as key to the integration of ICT into mathematics teaching. It also suggested that much could be learned about the ways in which teachers make use of ICT in their teaching by looking at the interaction between teachers' use of ICT and their professional knowledge base for teaching as defined by Shulman (1987). The review of the literature on the types of knowledge employed by teachers in their teaching offered a way forward with the analysis of the data already collected. A theoretical perspective from which to account for the observed variety in response of mathematics teachers to using ICT was thus developed and its usefulness was then justified by using it to analyse the data already collected. The study was carried out in 1997.

TEACHERS' USE OF ICT

There is a substantial body of research on the use of ICT in mathematics, but most of it concentrates on pupils' understanding of mathematics when the new technology is being used. For example, Simmt (1997) observes that studies involving graphics calculators and computers tend to be focused on student learning. Such studies looked at the effect the use of ICT on students' learning (they are the learners!), whereas teachers were expected to attend a one or two day training course in the use of ICT and then implement it into their teaching.

However, it is equally important to look at teachers' learning experiences of doing mathematics supported by the use of ICT as this might give an insight into why teachers use or do not use ICT in their instruction and why they use it the way they do. For example, in an Australian-wide survey undertaken to determine the most appropriate and effective ways in which classroom teachers have acquired the necessary skills and knowledge to use technology in their curriculum, Sherwood's (1993) data shows some changes in pedagogical practices. These changes appear to be more influenced by the interaction of the teacher him/herself as an individual, over time, with the technology rather than being influenced by variables such as teacher education, national culture and policy, specific types of software or software-use philosophy.

Reading through the literature on teachers' adoption of the new technology (for example, Marcienkievicz, 1994; Mandinach, 1994), an interesting strand describes the stages that beginning or novice (in using ICT) teachers pass in coming to make effective use of computers in teaching and learning. Teachers' uptake of ICT is regarded as a phased-process and it is suggested that those teachers who teach in schools with a good provision of ICT [2], where there is a relatively easy and immediate access to it and who have been using ICT in their mathematics lessons for 3 years or more, are more likely to use ICT as an integral part of their teaching and are more likely to reflect critically upon the ways ICT has been used in their mathematics lessons. Critical reflection is found to be necessary in order to determine if their incorporation of technology enhances or undermines the integration of technology. There is no 'right way' of using the ICT. This is a long-term process that builds on the teachers' increasing expertise and willingness to use the technology in new ways, and on what they learn from their students in the classroom as they use computers. Such activity is central to the process of questioning the technology. Teachers have to search for appropriate ways of enhancing teaching and learning.

What the existing research lacks is a deeper investigation of what teachers do with the new technology when they use it. What types of knowledge do teachers use when planning for an ICT lesson or designing an ICT based mathematics activity? What do teachers say about how their teaching is being affected by ICT use? What about the changes the use of ICT bring to the subject they master and teach?

The study reported in this paper aims at testing empirically some of the existing insights which surfaced from the literature reviewed concerning teachers' use of ICT as well as identifying those factors most important in the study of the integration of ICT into the teaching of mathematics. Thus the intention was to involve in the study a number of secondary mathematics teachers who have been using the new technology in their teaching for a number of years and seek out their views about the changes/challenges the use of ICT had brought to their profession.

THE STUDY

Sampling

According to the findings of the literature concerning teachers' use of ICT mentioned above, the first step in identifying these teachers was to find those schools known for having good provision of IT and where teachers have been using IT in the teaching of mathematics for a number of years. The NCET mathematics and ICT database '*Curriculum IT Support for mathematics*' was thus consulted. The database comprises reports from surveys in schools across the UK. It was intended for secondary mathematics teachers in the UK who wanted to know about interesting ICT developments in other schools. Each school in the database gave a brief description of the good features of the use of ICT within their mathematics departments, as well as mentioning those mathematics topics where ICT was being used. For example, in the database questionnaire, each school ticked topics referring to the use of ICT for learning and teaching mathematics that they would be willing to discuss with visitors from other schools, and possibly demonstrate to them, such as: *Mathematics*, e.g., algebra, probability, modelling; *Hardware*, e.g., graphics calculators, desktops; *IT*, e.g., integration of IT into workscheme, resources developed in school; *Students*, e.g., Years 12 and 13; *Applications*, e.g., databases, internet, graph plotters, symbol manipulators; *Entitlement Opportunities*, e.g., learning from feedback, observing patterns, exploring data and *Computers*, e.g., PCs, Apple computers. At the time when I accessed the database, there were around 70 records, with three secondary schools in the same area.

The study consisted of two phases: the first phase included a questionnaire survey for collection of mainly quantitative data about the teachers in order to obtain some background information about them. The purpose of the second phase was to follow in more depth some of the issues arising in the first phase of the data gathering process. Qualitative methods of inquiry, in particular interviews, were used with a few of the respondents to the questionnaires who volunteered themselves.

The questionnaires

The first phase of data gathering was carried out through the use of a mail questionnaire. The three of the schools in the same geographical area [3] were approached by letters to the head of the mathematics department and to the teachers in the department. The purpose of my research was explained and teachers were asked to complete the brief questionnaire enclosed. Teachers were reassured of the confidentiality of their answers and asked to seal the envelope provided once they filled in the questionnaire.

The researcher was aware that, when using questionnaires, not only is there a loss in sample numbers but also the fact that the non-respondents may differ significantly from the respondents. In the case of this study it was expected that there would probably be more teachers not using ICT among the non-respondents than those using ICT. However, a high rate of non-responses was not considered to 'distort' the general picture of effective use of ICT in mathematics.

The questionnaires were mainly intended to gather some factual information on teachers' background and practical issues regarding their use of the new technology. These items were produced by the researcher from ideas raised in past surveys on the use of IT in schools (Byard, 1995, Downes, 1993). The teachers were asked to answer items on their mathematics teaching experience, computer experience, type of training received in the use of ICT, familiarity with ICT, personal expertise and expertise in teaching with the available tools in their schools as well as frequency of use of the new technology in the mathematics lessons. Teachers' views on the influence the use of ICT has had on their knowledge of the subject matter, on the teaching of mathematics and on their pupils' learning of mathematics were also sought. These questions were both closed and open-ended in nature. With the exception of the items intended to explore teachers' opinions about the factors that most influenced them to take up the use of ICT, as well as the advantages and disadvantages of doing so, all the remaining questions were of Likert type scales.

Out of the three schools where the questionnaires were sent, one school failed to send back any questionnaire and out of the 22 mathematics teachers in the other two schools, 10 replied to the survey. Four out of the nine mathematics teachers from school A and six out of the 13 mathematics teachers from school B filled in the questionnaires.

The intention of the questionnaire was to help teachers reflect on some issues concerning their use of the new technology by completing the questionnaire, issues which would be investigated in more depth in the follow-up interview. Detailed analysis of the questionnaire survey was not appropriate (due to the lower response rate and the self-selected nature of the respondents) but the summary below gives a general picture of the situation.

All the 10 mathematics teachers who replied to the questionnaire had been using ICT in their teaching for more than three years, had been teaching in schools where there has been a relatively easy and immediate access to it and regarded themselves confident and competent in using the new technology. All the teachers were familiar with both computers and graphics calculators and they stated that they received adequate training through more than one type of course. Teachers' answers to the questions suggested that they viewed themselves as being knowledgeable in using the new technology for personal use as well as in teaching with it. They all agreed (and six strongly agreed) that their own mathematics, as well as their teaching of mathematics, had been enriched by the use of the new technology. Especially with these last questions, teachers were asked to tick the appropriate box as well as to expand on their choices of answers. None of the teachers expanded on the choices they made. However, it was hoped that at least teachers were provoked to reflect on such questions, which were going to be addressed again in the interviews.

The interviews

In order to investigate in more depth the use of the new technology by teachers in their teaching, interviews were carried out with seven teachers, six females and one male (TA1, TA2 from school A and TB1, TB2, TB3, TB4, TB5 from school B) who agreed and found the time to be interviewed. Using structured interviews, researchers usually remain on the surface level of teachers' views on their use of ICT, on the same level as a good questionnaire. However, it was felt that the narrative mode of the interview would allow the researcher to inquire in more depth teachers' views on whether and how they believed the availability and use of the new technology affected their mathematics and the school mathematics. It would also outline teachers' views on how they believed their teaching has changed as a consequence of reflecting on their experience at integrating ICT in their lessons. It was also hoped that prompting teachers to illustrate their answers with specific examples as well as asking supplementary questions (such as *What do you mean by...? Can you tell me some more about...?*) would help with clarifying points as well as with eliciting further detail about specific issues. The intention was to find out these teachers' views with regard to their experience of using IT in their teaching practice and the meanings they attached to certain events that took place in their mathematics classes in order to identify factors most important in their integration of ICT. The interviews were recorded on audio-tape.

The semi-structured nature of the interviews yielded the following categories of teachers' answers regarding their use of ICT: types of hardware available in schools and accessibility, software & resources on the use of ICT and the departmental policy regarding ICT use. Other responses gave an account of teachers' reasons for starting to use ICT, reasons for using ICT in their mathematics lessons as well as reporting on teachers' views regarding the value of using ICT with respect to today's school mathematics, teaching of mathematics and teachers' knowledge of the subject matter they teach.

DATA ANALYSIS

In the following I briefly discuss the categories of answers from both the questionnaires and the interviews in order to provide a broad picture of the schools and the teachers involved in this study.

Types of hardware available in schools and accessibility

Both schools have a very good provision of computers and graphics calculators available to teachers and pupils to make regular use of them in their subject area. In school A, the two teachers have in their classrooms: two computers, two lap-tops and a set of graphics calculators which are readily and easily accessible. In school B there is one computer in each classroom and five computers which can be wheeled whenever necessary. There are also four different network rooms for use by whole classrooms and a number of graphics calculators used as a resource especially by Y7,

Y8 and Sixth form. In both schools there are several items of educational software in the mathematics department (see below), as well as resources (books, worksheets) on how to use them.

Software & resources on the use of ICT

School A operates a resource-based, individualised scheme of learning, thus the teachers encouraged the use of SMILE activities on a regular basis. Other pieces of software used in the mathematics department are LOGO, database (Grasshopper), spreadsheets and Geometry Inventor. A lot of the materials on how teachers should use these applications with pupils in the mathematics lessons are written by staff at the school. If materials do not exist for a particular application, then teachers say they find it difficult to use it and tend not to use the application too often "the more able pupils use Geometry Inventor... not enough materials, thus not very easy to use" [TA1]. In school B, teachers use the Nuffield Advanced Mathematics syllabus that is geared towards the use of ICT. The syllabus consists of activities designed such that pupils can use their graphics calculators.

Materials on using ICT in mathematics lessons come from training courses, INSET courses or "from very good departmental INSETs", as the head of department put it, where teachers share practice. These materials, some of which being produced by the teachers in the department, help teachers use the existing educational software such as SMILE, LOGO, Geometry Inventor, spreadsheets and database programmes in their teaching.

Departmental policy

The analysis of the interviews suggests that all the teachers interviewed are enthusiastic about the new technology and, to varying degrees, well into using it. Their uptake and use of ICT have been well supported and in both schools there are clear guidelines on how ICT should be used. The teachers follow closely the departmental policy which has definitely contributed to these teachers' uptake and use of IT in their teaching. While some teachers use ICT because they think is a better approach to teaching a piece of mathematics, other teachers use ICT because they have a syllabus to follow which recommends using, for example, LOGO with pupils for one day a week. By following the departmental guidelines, this is how some teachers feel they try to "... make the use of computers part of the mathematics course". It was not possible to draw a clear conclusion about whether the use of IT is regular part of these teachers' practice, but the departmental policy indicates that the use of certain pieces of software is time-tabled at certain dates and for certain periods of time during the academic year. While this ensures that pupils benefit from learning mathematics with IT, it may not encourage teachers to think of using IT as regular part of their teaching or to explore the potential of using IT in all areas of mathematics.

Teachers' teaching of mathematics with ICT

All the teachers interviewed seemed to believe that the new technology could have a great impact on the teaching and learning of mathematics. The two young teachers teaching in school A did not seem to doubt the benefit of using the new technology. They were enthusiastic about using it in their mathematics lessons, questioning themselves more on *when to use it*, "I think it's to know when to set students that kind of work" [TA1] rather than *why use it*. This might be a consequences of the recent developments in initial teacher training courses which are more concerned than a few years ago with training student teachers to use ICT across the curriculum and in their main subject (Department for Education and Employment, 1998). Young teachers are thus more likely to have positive attitudes towards the use of ICT and to be in a better position to implement ICT in their teaching if they witness their lecturers modelling the use of ICT (Byard, 1995) and if they themselves witness the power of ICT in their own learning.

Teachers witnessed changes related to their teaching styles since their uptake of IT. The use of IT enabled the teachers to be more flexible in their teaching approach. For example, teacher TA2 felt that her teaching has changed since her pupils began using computers. She claims that the use of computers frees her to move around and observe her pupils doing mathematics more than usual. More emphasis is on her role as facilitator. The teacher is still the teacher but "there are going to be so many other 'teachers' in the room" [TB2]. On the other hand, TB1, with 20 years of teaching experience, feels that anything that comes up on the computer she has to point out to them and she also expects them to find the mathematical principles behind what they get on the computer because otherwise she would not be a mathematics teacher, "For me, I do regard myself as a mathematics teacher rather than an IT teacher".

These teachers also feel that the availability and use of IT has changed the A-level mathematics curriculum and believe that it is graphics capabilities that make a difference to the way pupils conceptualise and understand mathematics, "The breadth [of school mathematics] has changed as well as the ways mathematics is understood, learnt and taught" [TA1]. Also, the use of IT has led to less of a difference between KS3 (11-14 year old pupils) and KS4 (14-16 year old pupils), thus the hierarchy and order in which mathematics concepts could be acquired changes: "pupils do higher level mathematics"[TA1].

Teachers' views regarding the value of using ICT with respect to their mathematics

When filling in the questionnaires, the teachers agreed that the new technology enriched their knowledge of the subject matter (seven - strongly agree, three - agree). In the interviews, some teachers expanded on how they felt ICT had enriched their mathematics. For teacher TB5, who teaches 13-18 year olds, the use of ICT "...helps me with further maths – plotting curves". This teacher regards the graphical calculators invaluable in teaching graphical techniques at both GCSE and A-level.

Another teacher, TB3, started to explore fractals with LOGO and now presents his work at conferences. Being able to do things more accurately and at speed is regarded by these teachers as a way in which ICT enriches their mathematics work. One teacher states that she uses spreadsheets for "speed more than anything else, more organising; is helping me do things faster. This is how I feel IT enriches my maths other than just support what we are doing" [TA1].

Some teachers however feel that the new technology cannot possibly affect the mathematics of which they already have a good understanding, "I have a very good understanding of equations. IT does not really challenge my understanding of maths", but it enriches the mathematics they did not understand before using ICT (such as different algorithms, recursive procedures, geometry or anything they learnt by rote) which are now clearer to them, more intuitive. For example, TA2 believes that "... [Geometry Inventor] really made my geometry much more intuitive ...thinking about, rather than like a whole lot of facts which is how I did it at school" whereas TA1 states that "spreadsheets help me see the algorithms, really see the mathematics working, in action". These teachers' other answers given in the questionnaires and interviews also suggest that using ICT to perform algorithms and numerical calculations or to teach recursive procedures and geometry is very common in their mathematics lessons.

The teachers were also asked their opinion on whether or not they felt the use of ICT has challenged their mathematics. Although most of the teachers ticked the 'yes' box, very few teachers expanded on their answers to this particular question in the interview. For example, one teacher feels that her knowledge of school mathematics is not challenged, whereas when she uses ICT for herself "that it really actually challenges my mathematics" [TB1].

To summarise, the study reported here provided data on a small number of teachers who have been using ICT in their teaching for more than 3 years and regard themselves as competent and confident when doing so. These teachers' views regarding the incorporation of ICT in their mathematics lessons as well as their views regarding the value of using ICT with respect to the teaching of mathematics and their knowledge of mathematics were sought and presented. While it provided me with no fundamentally new insights into teachers' use of ICT in mathematics, the study served as input into the search for a new and interesting direction of research. The data collected suggests that teachers feel that their understanding of mathematics and teaching of mathematics is enriched and challenged as a result of using ICT. But how can I, the researcher, account for what teachers say about the ways their teaching of mathematics is being affected by ICT use and the changes the use of ICT bring to the subject they master and teach? I felt that a review of the research concerned with the wealth of knowledge employed by teachers in their teaching would be relevant. Thus, in the following, a number of such are reviewed.

TEACHERS' PROFESSIONAL KNOWLEDGE BASE FOR TEACHING

Shulman once described the teaching profession as "Those who understand, teach" (Shulman, 1986, p.5). But what do teachers need to understand in order to teach? Researchers have attempted to investigate the professional knowledge base for teaching from a variety of perspectives. Wilson, Shulman and Richert (1987) proposed a model of the components of the professional knowledge base for teaching described as being "the body of understanding, knowledge, skills and dispositions that a teacher needs to perform effectively in a teaching situation" (p.106). They found that teachers, in the course of preparing lessons and teaching, draw upon many types of knowledge when making decisions about the content of their lessons. Teachers obviously make use of the extent and depth of knowledge of the subject they teach, called subject matter_knowledge in this model. Many other researchers have explored teachers' subject matter knowledge and the role it plays in teaching (Shulman, 1986, Even, 1990). Wilson *et al.* (1987) also pointed to the importance of teachers having curricular knowledge (knowledge of instructional materials, of programs designed for teaching a particular topic, the materia media). Another component of the professional knowledge base for teaching is knowledge of other subjects and how they relate to the subject taught. These researchers also proposed including in this model teachers' knowledge of learners (knowledge of group dynamics, behaviour, how they interact and co-operate, responsiveness to learning tasks) and, more generally, knowledge of school context (of other teachers, of classrooms, of teaching resources, of departmental policies, of school ethos). Knowledge of education plays an important role as it provides teachers with the means to construe and interpret classroom experiences and to reflect on them. It was also proposed that researchers should approached the study of the teaching profession by focusing on teachers' general pedagogical knowledge which is knowledge of pedagogical principles and techniques that are not bound by topic or subject matter, and more specialised pedagogical knowledge, namely pedagogical content knowledge. This last component is enhanced by all the types of knowledge mentioned above and represents knowledge about how to teach mathematics and how to transform and represent mathematics ideas and topics to pupils.

In the following I focus on the two knowledge domains most central to the professional knowledge base for teaching, namely subject matter knowledge (SMK) and pedagogical content knowledge (PCK).

SMK consists of knowledge of the extent and depth of the subject, the structure, knowledge of procedures and strategies (Shulman, 1986), whereas pedagogical content knowledge (PCK) represents, in the context of mathematics, knowledge of how to teach mathematics, how to transform and represent mathematics ideas and topics to pupils.

> Within the category of pedagogical content knowledge I include, for the most regularly taught topics in one's subject area, the most useful forms of representations of those

ideas, the most powerful analogies, illustrations, examples, explanations and demonstrations - in a word, the ways of representing and formulating the subject that make it comprehensible to others.... (Shulman, 1986, p.9)

Shulman also included in this definition of PCK teachers' knowledge of pupils, their understanding of conceptions and preconceptions that students of different ages and backgrounds bring with them to the learning.

Research evidence indicates that subject matter expertise influences the pedagogical reasoning of a teacher. Studies show that teachers with limited SMK transform their SMK poorly (Smith and Neale, 1989), whereas a deep SMK could lead to teachers accommodating new ideas, new pedagogical conceptions, despite the traditional approach they adopted before the innovation programmes was introduced (Lloyd and Wilson, 1998). However, very good SMK does not necessarily mean good PCK (Grossman, Wilson and Shulman, 1989). PCK is more than this; it involves the transformation of SMK. Researchers have used Shulman's characterisation of pedagogical content knowledge productively, but there are difficulties with using it in practice. Findings of a significant number of studies on effective teaching have resulted in a widening of the definition of SMK. Researchers have long argued that the ability to represent the subject matter is an important aspect of an individual's PCK. To facilitate the development of powerful, appropriate representation, teachers need to evaluate their own understanding of the subject matter they teach. They must generate representations that take into account their understanding, as well as knowledge that is already held by their pupils. Teachers hold specific, favoured representations of particular ideas for their own purpose. When teaching, they further develop the capacity to introduce alternative representations of the SMK. Teachers actively create these representations.

Researchers such as Ernest (1989) suggested that subject matter needed for teaching mathematics should include knowledge of mathematics as well as knowledge about the nature of mathematics and understanding of what it means to know and do mathematics. Even (1990) criticised the research in teachers' SMK area as being too general and not topic-specific and suggested that teachers should have knowledge and understanding of the concept to be taught, essential features of this concept, different representations and alternative approaches. Grossman *et al.*, (1989), Ernest (1989) and Thompson (1992) suggested including here teachers' beliefs about the subject matter they teach as these contribute to the ways in which teachers think about the subject they teach and the choices they make in their teaching.

Transformation of subject matter for teaching (Shulman, 1986) occurs as the teacher *critically reflects on* and *interprets* the subject matter; finds multiple ways to *represent* the information as analogies, metaphors, examples, problems, demonstrations and classroom activities, *adapts* the material to students' abilities, gender, prior knowledge and finally *tailors* the material to those specific students to whom the information will be taught. As students are multiple, so representations must be various. Thus teachers should possess a representational repertoire for the subject matter they teach.

McNamara (1991) and Marks (1991) have pointed out the difficulty of making a clear distinction between SMK and PCK as it all depends on how statements about mathematics teaching and learning are interpreted. Teachers cannot, for example, always articulate how subject matter shapes their teaching practice even though they think it is important. Different teachers incorporate subject matter and pedagogic content knowledge in their teaching in different ways.

The sources for a knowledge base for teaching mentioned in this section are plentiful and the relationship between them very complex. In Shulman's model, these components float on the page, as the relationships between themselves are too complex to pin them down. However, such simplification is needed when conducting a study as an intermediary step towards the quest for general principles of effective teaching.

A POSSIBLE THEORETICAL FRAMEWORK

Adopting a theoretical perspective which will take into consideration some of the categories of the professional knowledge base for teaching (SMK, PCK and the interaction between them) is thought to be a way forward with the analysis of the data already collected. In order to justify the usefulness of such perspective, I will revisit some of data collected in this study.

For example, teacher TA2 believes that "... [Geometry Inventor] really made my geometry much more intuitive ...thinking about, rather than like a whole lot of facts which is how I did it at school" whereas teacher TA1 states that "spreadsheets help me see the algorithms, really see the mathematics working, in action". These views seem to suggest that the use of ICT helped these teachers better understand the mathematics they did not understand very well or they learnt by rote such as algorithms, recursive procedures, even geometry. These two teachers also said that using ICT to perform algorithms and numerical calculations or to teach recursive procedures and geometry is very common in their mathematics lessons. Thus teachers reflect on their experiences as learners of mathematics and try to find better approaches to teaching those topics they did not understand very well. These teachers found that their learning experiences of doing mathematics with ICT enriched their representational repertoires with 'better ways' of teaching some areas of mathematics than they usually used in their instruction.

For teacher TB5, the use of ICT "... helps me with further maths – plotting curves". In this teacher's case, the graphics capabilities of the new technology enriched her understanding of further mathematics through the new, visual representations enabled by it.

Another teacher started to explore fractals with LOGO and now presents his work at conferences. His SMK has thus been enriched by adding to it a new mathematics topic. The extension of this teacher's mathematical knowledge was enabled by his experience of doing mathematics supported by ICT.

TB1 stated that she felt that her knowledge of school mathematics was not challenged, whereas when she used ICT for herself "that it really actually challenges my mathematics" [TB1]. It is hard to know exactly what this teacher meant, since it was not investigated further when the interview was carried out. But could this view indicate that this teacher's knowledge of mathematics, her ideas about how to transform and teach the topic are in fact challenged when ICT is being used?

THE NEXT STEP

This paper reported on an exploratory study in which the researcher embarked on a journey from a theory free situation to the development of a theory. This theory will be further developed and refined in further work, the longer term intention being that of developing a theoretical model which accounts for teachers' learning about ICT and teachers' incorporation of ICT into their planning for teaching mathematics. As mentioned above, the model will take into consideration some the categories of the professional knowledge base for teaching identified by Shulman and will aim at investigating the interaction between these components when ICT is used. It is hoped that this investigation will throw light on how teachers' knowledge of the subject matter they teach as well as their pedagogical content knowledge (Shulman, 1987) are affected by and affect the use of ICT. Such knowledge could only contribute to the successful implementation of the new technology in schools.

NOTES

1. Throughout this paper, terms such as the new technology, IT and ICT are used interchangeably in order to refer to graphics calculators and computers together with applications such as spreadsheets, graphics packages, computer algebra and geometry systems.

2. Good provision of ICT consists of a sufficient number of ICT tools, educational software for the teaching and learning of a subject area and materials on how to use them, which are easily and readily available to teachers and pupils when needed.

3. As the Curriculum IT Support for Mathematics database can be accessed on the Internet, there is no mention of the name of the city/area where the study was conducted in order to protect the identity of the schools.

REFERENCES

Byard, M. J.: 1995, 'IT under school-based policies for Initial Teacher Training.' *Journal of Computer Assisted Learning, 11*, 128-140.

Department for Education and Employment: 1998, *Circular number 4/98: Requirements for Courses of Initial Teacher Training*. London: DfEE.

Downes, T.: 1993, 'Student-teachers' experiences in using computers during teaching practices.' *Journal of Computer Assisted Learning, 9*, 17-33.

Ernest, P.: 1989, 'The knowledge, beliefs and attitudes of the maths teacher: A model.' *Journal of Education for Teaching, 15*(1), 13-33.

Even, R.: 1990, 'Subject matter knowledge for teaching and the case of functions.' *Educational Studies in Mathematics, 21*, 521-544.

Grossman, P. L., Wilson, S. M. and Shulman, L. S.: 1989, 'Teachers of substance: Subject matter knowledge of teaching.' In M. C. Reynolds (ed.), *Knowledge Base for the Beginning Teacher* (pp. 23-36). New York: Pergamon Press.

Lloyd, G. and Wilson, M.: 1998, 'Supporting innovation: The impact of a teacher's conceptions of functions on his implementation of a reform curriculum.' *Journal of Research in Mathematics Education, 29*(3), 248-274.

Mandinach, E. B. and Cline, H.: 1994, *Classroom Dynamics: Implementing a Technology-Based Learning Environment.* Hillsdale, New Jersey: Lawrence Erlbaum Associates.

Marcinkiewicz, H.R.: 1994, 'Computers and teachers: Factors influencing computers use in the classroom.' *Journal of Research on Computing in Education, 26*(2), 230-235.

Marks, R.: 1991, 'Pedagogical content knowledge: from a mathematical case to a modified conception.' *Journal of Teacher Education, 41*(3), 3-11.

McNamara, D.: 1991, 'Subject knowledge and its application: Problems and possibilities for teacher educators.' *Journal of Education for Teaching, 17*(2), 113-128.

Simmt, E.: 1997, 'Graphing calculators in high school mathematics.' *Journal of Computers in Mathematics and Science Teaching, 16*(2/3), 269-289.

Sherwood, C.: 1993, 'Australian experiences with the effective classroom integration of information technology: Implications for teacher education.' *Journal of Information Technology for Teacher Education, 2*(2), 167-179.

Shulman, L. S.: 1986, 'Those who understand: Knowledge growth in teaching.' *Educational Research, 15*(2), 4-14.

Shulman, L. S.: 1987, 'Knowledge and teaching: Foundations of the new reform.' *Harvard Educational Review, 57*(1), 1-22.

Smith, D. C. and Neale D. C.: 1989, 'The construction of subject matter knowledge in primary science teaching.' *Teaching and Teacher Education, 5*(1), 1-20.

Thompson, A. G.: 1992, 'Teachers' beliefs and conceptions: A synthesis of the research.' In D.A. Grouws (ed.), *Handbook of Research on Mathematics Teaching and Learning* (pp.127-146). New York: Macmillan Publishing Company.

Wilson, S. M., Shulman, L. S. and Richert, A. E.: 1987, '150 Different Ways of knowing: Representations of knowledge in teaching.' In J. Calderhead (ed.), *Exploring teachers' thinking*, (pp. 104-124). London: Cassell Educational Limited

6   MODELS AND SUPER-MODELS: WAYS OF THINKING ABOUT PROFESSIONAL KNOWLEDGE IN MATHEMATICS TEACHING[1]

Stephanie Prestage and Pat Perks

School of Education, University of Birmingham

Teaching about teaching is a complex process requiring knowledge about teaching as well as knowledge about teaching about teaching. We have published findings on research carried out over the last few years about teachers' subject knowledge. This research led to the proposal of a model for thinking about subject knowledge which distinguishes between knowledge needed to pass an examination and knowledge needed to help someone else to come to know that knowledge. The first is necessary but not sufficient for the latter. This model for thinking about subject knowledge has led to proposals for similar models for thinking about other aspects of teacher knowledge and has more recently developed into a parallel model for thinking about teacher education.

INTRODUCTION

Teaching about teaching is a complex process mainly because teaching itself is a complex profession. The teacher is required to hold knowledge about pupils, systems and structures; knowledge about styles of teaching and learning; knowledge about management, resources and assessment as well as knowledge about the subject. Existing research in the area of teachers' knowledge offers definitions of professional knowledge as well as explanations for the different forms of knowledge that a teacher holds, (Brown and McIntyre, 1993; Cooper and McIntyre, 1996; Desforges and McNamara, 1979; Ernest, 1989; Marks, 1990; Calderhead and Shorrock, 1997; Banks, Leach and Moon, 1999).

> It is clear that learning to teach involves more than a mastery of a limited set of competencies. It is a complex process. It is also a lengthy process, extending for most teachers well after their initial training. (Calderhead and Shorrock, 1997, p.194)

The first part of this paper considers teachers' subject knowledge as it is from research in this area that a model describing aspects of a teacher's knowledge about mathematics emerged. Subject knowledge is an aspect of teachers' professional knowledge known to be problematic, but one which "has provoked more controversy than study" (Grossman, Wilson and Shulman, 1989).

> While one can infer from studies of teacher thinking that teachers have knowledge of their students, of their curriculum, of the learning process that is used to make decisions, it remains unclear what teachers know about their subject matter. (Wilson, Shulman and Richert, 1987, p.108)

Research in the particular area of subject knowledge and pedagogical content knowledge (Shulman, 1986; Wilson *et al.*; 1987, Tamir, 1988; Aubrey, 1997) explores the transformation of subject matter knowledge for the classroom, *viz.* teachers' knowledge about explanations, tasks and activities, about styles of teaching and learning but does not include explicit detail of how such subject knowledge is held in an intellectual way by teachers other than is demonstrated by the activities chosen or the explanations given. The mathematics is rarely explicit in these descriptions. Whilst Shulman and others have categorised the different components of subject knowledge and discussed its transformation through classroom events, our research data was used to investigate the ways in which teachers' subject knowledge in mathematics is held and transformed.

Beliefs about teaching lead to us to agree with Buchmann (1984) that teachers need a rich and deep understanding of their subject in order to respond to all aspects of pupils' needs: "Content knowledge of this kind encourages the mobility of teacher conceptions and yields knowledge in the form of multiple and fluid conceptions" (p.46). We take subject knowledge to be knowledge about the subject matter in mathematics, knowledge about its structure, the body of concepts, facts, skills and definitions, as well as methods of justification and proof. Through evidence from the research a hypothesis was developed that teachers' subject knowledge in mathematics might be held in two forms either as *learner-knowledge* in mathematics or as *teacher-knowledge* in mathematics, the former is the knowledge needed to pass examinations, to find solutions to mathematical problems; the latter is the knowledge needed to plan for others to come to learn the mathematics. Auditing the former might be necessary (currently demanded on Initial Teacher Education (ITE) courses) but is not sufficient for the developing teacher. Such an audit would offer a limited list-like perspective of the knowledge held by a student teacher, a view shared by other authors.

> The shared assumption underlying [such] research is that a teacher's knowledge of the subject matter can be treated as a list-like collection of individual propositions readily sampled and measured by standardised tests ...[these] researchers ask what a teacher knows and not how that knowledge is organised, justified or validated ... [such research] has failed to provide insight into the character of the knowledge held by students and teachers and the ways in which that knowledge is developed, enriched and used in classrooms. (Wilson *et al.*, 1987, p. 107)

By proposing a model for subject knowledge, we offer a structure for thinking about its complexity in the teaching of mathematics. It is the very 'character of the knowledge' in mathematics that we sought to provide insight.

METHODS AND ANALYSIS

The nature of the initial research was based within an interpretive research paradigm (c.f. Bassey, 1995). Existing research was used to develop ideas and to provide a basis for explanation. The data reported here is a sub-set of a wider set of data exploring aspects of professional knowledge, based on case studies which included interviews with primary, secondary (over two years, Prestage, 1999) and pre-service teachers (over one year, Perks, 1997). The experienced teachers were subject leaders nominated for their expertise by advisors to join a National Curriculum project (Prestage, 1996). Expectations of this group in terms of their individual understanding of subject knowledge were, legitimately, high. The pre-service teachers were following a one-year postgraduate course in education, with the majority having a first degree in mathematics. Subject knowledge was explored through questions about planning for teaching, from the interviews that took place with the expert teachers and from the data from lesson planning with the pre-service teachers.

The literature on subject knowledge provided a context and a springboard for the work but did not offer a way forward with the analysis of the data from the experienced teachers (Prestage, 1999). The analysis looked to the wider debate on the professional development of teachers and in particular to Eraut's (1994) comment on the twofold problem for professional education.

> First, certain systems of thought or paradigms dominate a profession's thinking in a way that they are passed unquestioned from one generation to the next ... The second problem is the converse of the first. To make practical use of concepts and ideas other than those embedded in well-established professional traditions requires intellectual effort and an encouraging work-context. The meaning of a new idea has to be rediscovered in the practical situation, and the implication for action thought through. (p.49)

Eraut's quote was crucial to the analysis. He presents the idea that teachers acquire knowledge through professional traditions and in order to accommodate new ideas will need to work on them in the classroom (practical wisdom) and then think through the implication of these actions (deliberate reflection). Whilst the quote from Eraut encompasses the wide range of knowledge gained by teachers, these ideas were used to classify the subject knowledge of teachers. Initially three phases – professional traditions, practical wisdom, deliberate reflection were defined to classify teachers' responses to questions about subject knowledge (Prestage, 1999, pp. 37-39).

1 Professional Traditions: This phase represents teachers accounting for their subject knowledge through reference mainly to the unquestioning acquisition of subject knowledge from the professional traditions of their own learning, through use departmental schemes, texts and government policies, of doing things because "that is the way they are always done".

2 Practical Wisdom: This phase represents teachers accounting for their subject knowledge through its rediscovering in the practical situation i.e. in the classroom as a consequence of observing and working with pupils. There is varied evidence in the research to suggest that subject knowledge may or may not develop as a consequence of teaching. This classification is used when teachers talk about altering aspects of their subject knowledge as a consequence of classroom outcomes.

3 Deliberate Reflection: This phase represents an explicit intellectual re-working by teachers of their subject knowledge beyond that gained from classroom practice. This classification acknowledges that more than practice might be used to develop subject knowledge and infers a deliberate 'standing back' from the classroom situation in order to re-think aspects of subject knowledge to plan for teaching.

The word phase was used to indicate the possibility of movement over time as a teacher's subject knowledge develops and transforms. For example if a teacher were asked about division of fractions:

- a professional tradition response might be based within the learner-knowledge i.e. to teach the algorithm 'change the sign to multiply and turn the second fraction upside down';

- a practical wisdom response might be to work in the classroom with a lot of activities that would help to demonstrate the algorithm;

- a deliberate reflection response might be an explicit integration of the mathematics behind the algorithm (reciprocal, inverse, generalisation) and be able to give the 'why' of the classroom activities chosen.

By analysing the teachers' responses to discussions of subject matter it was possible to categorise the phases in which some aspects of their mathematics knowledge was held. This analysis offered a way of explaining the difference in nature of the types of responses and how transition between the phases might be described. A model emerged that summarised our findings. The later part of this paper offers an extension of this model to the many different aspects of teacher's professional knowledge. These models in their turn offer a description of the variety of learner-knowledge which is needed by the teacher-educator and suggested the creation of a super model.

Whilst the roots of the model for subject knowledge lie in a variety of research, from which we offer exemplars, an explanation of its components is more easily given via our ITE students.

THE BEGINNINGS OF A MODEL: LEARNER-KNOWLEDGE AND PROFESSIONAL TRADITIONS

Graduate mathematician students of our secondary pre-service course naturally hold subject knowledge as learner-knowledge. They arrive with a certain amount of personal subject knowledge (*learner-knowledge*) that enables them to answer mathematical questions. When asked to calculate the division of one fraction by another, to differentiate a function, to solve a set of equations, all respond correctly. When asked why the answers are correct they do not know. They can do mathematics but they do not necessarily hold 'multiple and fluid conceptions'. ITE students also bring with them their personal beliefs and certain characteristics of 'being a teacher'. The view of teaching for these pre-service teachers is to replicate the learner-knowledge they hold for others to learn. Their subject knowledge is often ill-connected and they have to work on this when planning for teaching (Perks and Prestage, 1994). However, this interpretation of teaching is also held by some experienced teachers. The two quotes given are from heads of department:

I don't think it matters its just the order we do them in …I suppose the order I do things in is the order that I was taught, the order I have always done them I don't think it would have occurred to me [to change the order]

The order I did it in was how I was taught …angles on a straight line, angles at a point, then vertically opposite angles and then parallel line theorems, angles in polygons …

During their ITE course students gain different knowledge and understandings of other professional traditions – some national like the National Curriculum, the Numeracy Strategy, the examination system, each with their attendant exemplar materials, and some local traditions gained from particular school settings such as schemes and textbooks and the ways in which national policies are translated in different settings. Learner-knowledge and professional traditions merge in the first instance to create classroom events for others to engage with learning mathematics.

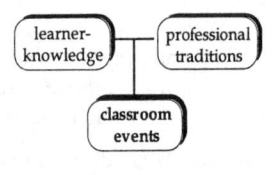

Figure 1

But this combination is also evident from the experienced teacher. High on the list when justifying decisions about the curriculum were text books and other departmental resources, experiences of learning mathematics, ideas related to teaching practices (Baturo and Nason, 1996; Ball, 1988; 1990) and from the new legislative curriculum:

It is perhaps based on two influences, how I was taught and the other… using a particular published scheme which has been the case in the past and again that sort of hierarchy is laid out. …. I'm teaching nets tomorrow. Why? It was put into the scheme of work by my predecessor

The content and the ideas for activities have broadly come from a selection of ideas in the teacher's book, from the maths scheme that we use. Because at the moment we don't have a maths policy as such, we don't have any maths guidelines and as a stop gap our maths is organised around Cambridge maths.

In my experience I have found that the SMP [secondary] books tend to jump around a lot rather than follow the progression throughout the National Curriculum levels in shape.. and [in particular] … as for rotational symmetry that comes much later on [in the scheme]. This is an area that I am intending to write a scheme of work around to follow the levels of National Curriculum more closely.

Additionally, the power of the new imposed tradition was also apparent from the data. The experienced teachers were asked about 'generalisation', an aspect of the curriculum not explicit in their schooling. Responses ranged from one secondary head of department saying that she did not know about it, did not teach it and that her department did not really do enough of it; another with 'we were never taught generalisation'. Progression was offered in terms of getting better at spotting patterns but no mathematical explanation was given for what 'getting better' meant. Justifications for classroom decisions were based on 'coursework activities in year 10 and 11' (Prestage, 1999, pp. 207-215). The teachers had no explicit learner-knowledge for 'generalisation' and could not see the connections to their own learning experiences where mathematical generalisations were numerous. Perhaps learner-knowledge is only accessible in its labelled form.

PRACTICAL WISDOM

There comes a moment for many teachers (often early in their professional education) when they realise that giving their learner-knowledge directly to pupils does not work:

> if you present a problem to the class and there is a need for them to know something about a particular shape and the area of it, and therefore they would set the agenda ... things are encompassed in a problem and the pupils are setting the agenda.

Reflection upon these classroom events, with the integration of learner-knowledge and professional traditions, leads to the beginnings of some practical wisdom that enables the teachers to adapt activities from the professional traditions to suit their particular circumstances.

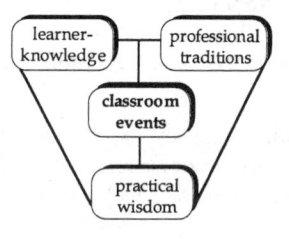

Figure 2

The pre-service teachers develop such practical wisdom during their teaching experiences as shown in their evaluation of lessons:

> Needed to spend longer than I thought going through triangles... even with further explanation, most of the pupils were still using slope and height to calculate area.... I tried explaining splitting triangle in two, drop a perpendicular..., still had a few problems... although by turning the triangle round to get long side on base (horizontal) seemed to be getting there. Needs reinforcing again.

Many of the experienced teachers talked in the interviews about the consequences of diverse classroom interactions, of altering the teaching decisions in order to respond to pupils' needs. One primary mathematics co-ordinator for whom the scheme was a dominant resource claimed that the use of texts was tempered by his classroom experience: "What I have actually discovered is that it is not really like that ... the timing is really important and the sequence is important."

Classroom experiences give rise to practical wisdom based on perceived learner needs and new explanations and contexts are found to support learner-knowledge. These classroom experiences are the ones most often described in the research. However, the learner-knowledge is not necessarily questioned, the teacher may still offer the rule "change the divide sign to a multiply and turn the second fraction upside down" but previous lessons may have included "how many quarters in one whole, how many in two etc." and other starting points to support the learning of the algorithm. Existing research makes assumptions that teachers have full access to subject matter knowledge and that this is transformed by the activities developed for teaching. We would argue that for both experienced and novice teachers much of this subject matter knowledge remains as learner-knowledge and is not transformed into teacher-knowledge, a finding supported by Aubrey (1997).

> There is however, little evidence to suggest that the development of project teachers' subject matter through teaching occurred. The capacity to transform personal understanding, thus, depends on what teachers bring to the classroom. Whilst knowledge of learning and teaching and classrooms increases with experience, knowledge of subject content does not. (pp.159-160)

TEACHER-KNOWLEDGE

In certain topics there was evidence from the research data that some teachers had thought more about the subject matter, beyond reacting to pupils. For one primary co-ordinator curriculum decisions were made in response to the pupils in her Y2 class "depending on where the conversation goes". There was evidence throughout the interview that she had made deliberate decisions about progression through the mathematics. For example, she was the only teacher interviewed who spoke positively about generalisation:

> This is the sort of thing I am looking for. [She picks up the generalisation card]. I am not entirely sure what it is so I had better start with a definition. I think it is about a general understanding about mathematical concepts. It is about a confidence it is about being able to use all sorts of mathematical concepts and matching them to the particular task that they theoretically (I hope) have chosen, ... but I realise it is what I choose most of the time and that is what I am heading towards all the time in any mathematical topic

This aspect of deliberate reflection towards teacher-knowledge was also partially evident in other interviews.

> Maybe I was fixed on number at one point and I have gone through a process with number and I think possibly over the last two years it has helped sitting down and thinking about progression and thinking about doing things as pupils need to know them

rather than in an order where I feel they need to know them and I haven't perhaps gone through that process with angle so therefore I am still thinking along a very set pattern.

This secondary head of department continued by describing how he decided to reject fractions as learner-knowledge for pupils despite existing professional traditions. He acknowledged that the time spent on the project had enabled him to question and challenge his learner-knowledge. This was not the experience of all the teachers on the project.

Teacher-knowledge also emerges occasionally in the student teachers. Jane had a lesson on constructing regular polygons, based on constructing isosceles triangles with the odd angle of 45°, these were cut out and reformed into regular octagons. Jane offered the objectives of the lesson from the practice of known skills to the development of the generalisation of the relationship with the angle at the centre of a circle, the angle in the triangle and the number of sides of the regular polygon. She was very aware of the problem of interference in this process of error in the construction and how this had to be related to number theory.

We believe therefore that 'good' teachers need to reflect upon classroom events not simply to consider their success or failure for the students but to reconsider their own personal understandings of mathematics, to reflect upon the 'why' not only of teaching but also of mathematics. We would argue that it is in this way that they come to own a better personal knowledge of mathematics (teacher-knowledge), that learner-knowledge (the only explicit content knowledge) requires transforming through deliberate reflection but that this analytic process requires a synthesis of the reflection on the three elements of figure 2. The model is completed in the form of a tetrahedron, where the struts represent the reflective/analytic process.

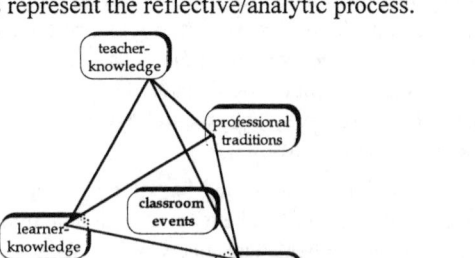

Figure 3

Shulman argued that pedagogical content knowledge was the missing paradigm for teaching and whilst we agree with his focus on the subject and its importance in teaching, we believe that designing activities for the classroom is necessary but not sufficient to develop the type of subject knowledge, the *teacher-knowledge*, we feel essential for the development of strong mathematics teaching. Neither is it acceptable to simply demand more learner-knowledge. More of this kind of knowledge is not sufficient for the development of teacher-knowledge. The analytic process has been long recognised and is essential for the intellectual professional.

> There is a long standing dilemma in teacher education. Ideally we may wish to have teachers who are not only competent actors in the classroom but also who are practitioners capable of understanding what they are doing, why they are doing it and how they might change their practice to suit changing curricula, contexts or circumstances. This produces a tension between the need for teachers to understand teaching and the need to be able to perform teaching. (Calderhead and Shorrock, 1997, p. 195)

Such teacher-knowledge in mathematics allows teachers not only to answer the questions correctly but also helps them to build a variety of connections and routes through that knowledge, that provides answers to *why* something is so (Prestage, 1999). It is our contention that only when such *teacher-knowledge* is informing classroom practice are the real needs of learners and the challenges of mathematics addressed.

PROFESSIONAL KNOWLEDGE

Our research to date has been mainly with subject knowledge. But imagine the model in figure 3 extended to include not only understanding and knowledge about mathematics but also about the other elements that the professional teacher needs to develop, issues related to assessment, class management, equal opportunities etc.. The pack of professional 'teacher-knowledge' cards might look like this:

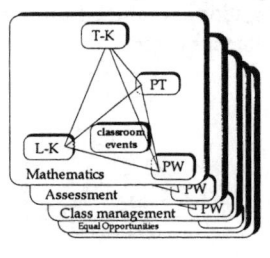

Figure 4

The complexity of the job is evident in this imagery. In terms of professional knowledge we are working with our pre-service teachers to develop not only teacher-knowledge in their subject matter but also in assessment issues, classroom management, equal opportunities and the like. ITE students need to gain understandings of the professional traditions in each of these cards, to gain practical wisdom across the range and to reflect upon these with their own learner-knowledge. Learner-knowledge might be based in the pre-service teachers' personal experiences, for example, much classroom management is based on their most memorable teacher but might lead to ill-thought through decisions in the moment. Professional traditions exist for each dimension, for example, public examinations in the assessment area.

Experience of the practical situation is vital to develop the practical wisdom which will inform about alternatives and appropriate actions. Teacher-knowledge requires knowledge in the three strands to be synthesised by reflection. Time is needed for these developments:

> People have to develop implicit theories of action in order to make professional life tolerable. There are too many variables to take into account all at once, so people develop routines and decision habits to keep mental effort at a reasonable level. To change the routine or question the theory is to reverse the process, to draw attention once more to myriads of additional variables, and to raise the possibility of paralysis from information overload and failing to cope. (Eraut, 1994, p. 34)

THE SUPER-MODEL

What then are the implications of this for thinking about a pedagogy for teacher education? Is there an equivalent teacher-knowledge which informs our practice? We believe that just as teaching mathematics needs fluid and connected knowledge of mathematics (teacher-knowledge) so too mathematics educators need an articulated, fluid and connected understanding of teaching mathematics education – the teacher-knowledge of mathematics education.

Firstly, mathematical subject knowledge for teaching about teaching mathematics. The teacher educator needs subject knowledge at the teacher-knowledge level. Most mathematics teacher educators have aspects of figure 5 in place (figure 3 repeated), learner-knowledge of mathematics, some knowledge of the professional traditions, a certain amount of practical wisdom combined through reflection to create teacher-knowledge for mathematics subject matter

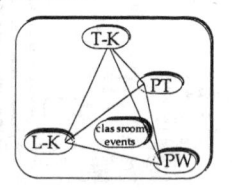

Figure 5

This mathematical learner-knowledge for a teacher educator then forms the basis, along with ITE professional traditions and the practical wisdom of teaching about teaching to create sessions for ITE students to work on **their** subject knowledge for teaching. Figure 5 lies at the learner-knowledge vertex in the following tetrahedron.

Professional traditions emerge from personal experiences, education and training, the current government teacher training policies, the mathematics and technology national curricula for teacher education (Department for Education and Employment, 1998) as well as inspection criteria against which judgements are made. Practical wisdom is gained from considering what the students need to know and how sessions might be constructed for them so that they engage in the ideas.

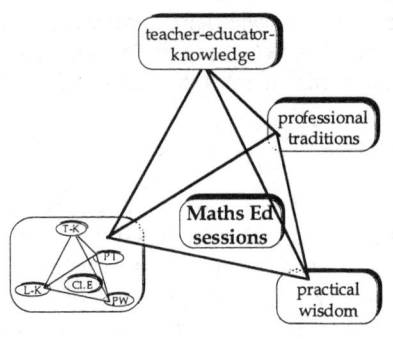

Figure 6

For those involved in teacher education the implications of this model extend to all other areas. The aim for our students should be that they begin to use their learning on the course for the whole of this pack of cards to abstract their knowledge to the teacher-knowledge level. The final model for teacher education thus integrates all of the cards from figure 4 as learner-knowledge:

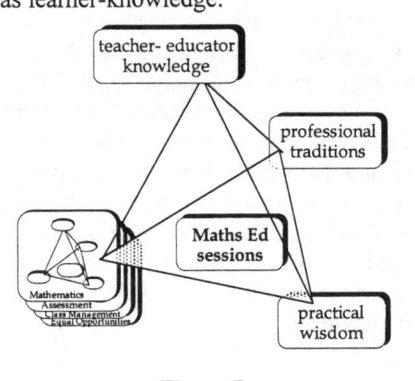

Figure 7

We have begun to explore these ideas elsewhere (Prestage and Perks, 1999a and b), analysing case studies of teacher education practice to determine the usefulness of the model. Looking at the nature of knowledge held for any of these cards has helped to identify where and why there is need for improvement in our own practice. Discussion from a BSRLM session (Prestage and Perks, 1999c) was wide ranging. One area centred on the thorny issue of the theory-practice divide in education. The diagram in Figure 7 offers a strong image to counter the idea of a theory-practice divide offering instead a theory-practice balance of the knowledge held by those involved in ITE. It was also suggested that the balance of practical wisdom, professional traditions and learner-knowledge might be weighted differently for each of the cards. For example in the area of equal opportunities, as women in education we have learner-knowledge and some practical wisdom but rely on professional traditions for other aspects of this wide field. Particular situations, particular contexts may not support explicitly all layers of professional development. This may depend on accessing a range of expertise (across the partnerships) at the teacher-knowledge level.

DISCUSSION

The reduction of teacher education to vocational competencies, audits and tests in core skills does not address the complex process of becoming a teacher. If teaching is a complex process (evidenced in the research) then, *sine qua non*, teaching about teaching is a complex process. The required audit of learner-knowledge offers no security that good teaching will follow – the focus on development is misdirected. Achieving high marks in a teacher numeracy tests offers no indication of my ability to teach my own subject nor integrate numeracy into my lessons – the attention to this type of detail is inappropriate.

Hence the models. The model for subject knowledge had helped us to articulate the importance of a fluid and connected subject knowledge for teaching which we now emphasise in our own teaching. The model for thinking about the process of teacher education might help to make explicit the multiple and fluid conceptions of the art of teaching mathematics education. Any description of the process of teacher development needs to be presented in a manageable form whilst at the same time presenting key elements. By using the struts of the tetrahedron to highlight the dynamic nature of reflection and the need for synthesis we hope to have presented a strong image for talking about aspects of teacher development. The model has offered a way of analysing aspects of our practice as teacher educators to aid its development. Once we have a shared understanding, a "wisdom of practice" (Shulman, 1986) we may be better prepared to move towards an articulation of a pedagogy for mathematics teacher education that begins to explain why learning about teaching is more than gaining craft knowledge.

NOTES

1. This paper is an expansion of the paper given at the BSRLM conference in November 1999.

REFERENCES

Aubrey, C.: 1997, *Mathematics Teaching in the Early Years: An investigation into teachers' subject knowledge*. Falmer Press: London.

Ball, D.L.: 1988, 'Unlearning to teach mathematics.' *For the Learning of Mathematics*, *8*(1), 40-42.

Ball, D.L.: 1990, 'Breaking with experience to teach mathematics: The role of a preservice methods course.' *For the Learning of Mathematics*, *10*(2), 10-16.

Banks, F., Leach J. and Moon B.: 1999, 'New understandings of teachers' pedagogic knowledge.' In J. Leach and B. Moon (eds.), *Learners and Pedagogy* (pp. 89-110). London: PCP.

Bassey, M.: 1995, *Creating Education Through Research*. England: Kirklington Moor Press (in conjunction with BERA).

Baturo, A. and Nason, R.: 1996, 'Student teachers' subject matter knowledge within the domain of area measurement.' *Educational Studies in Mathematics, 31*, 235-268.

Brown, S. and McIntyre, D.: 1993, *Making Sense of Teaching*. Buckingham: Open University Press.

Buchmann, M.: 1984, 'The priority of knowledge and understanding in teaching.' In L. Katz and J. Roth (eds.), *Advances in teacher education Volume 1* (pp. 29-50). New Jersey: Ablex.

Calderhead J. and Shorrock S. B.: 1997, *Understanding Teacher Education*. London: Falmer Press.

Cooper, P. and McIntyre, D.: 1996, *Effective Teaching and Learning: Teachers' and students' perspective*. Buckingham: Open University Press.

Department for Education and Employment: 1998, *Circular 4/98: Teaching: High Status, High Standards* (Annex G and Annex B). London: DfEE.

Desforges, C. and McNamara, D.: 1979, 'Theory and practice: Methodological procedures for the objectification of craft knowledge.' *British Journal of Teacher Education, 5*(2), 145-152.

Eraut, M.: 1994, *Developing Professional Knowledge and Competence*. London: Falmer Press.

Ernest, P.: 1989, 'The knowledge, beliefs and attitudes of the mathematics teacher: A model.' *Journal of Education for Teaching, 15*(1), 13-33.

Grossman, P. L., Wilson, S. M., and Shulman, L. S.: 1989, 'Teachers of substance: Subject matter knowledge for teaching.' In M. C. Reynolds (ed.) *Knowledge Base for the Beginning Teacher* (pp. 23-36). Oxford: Pergamon.

Hopkins, D. and Stern, D.: 1996, 'Quality teachers, quality schools: International perspectives and policy implications.' *Teaching and Teacher Education, 12*(5), 501-517.

Marks, P. R.: 1990, 'Pedagogical content knowledge: From mathematical case to a modified conception.' *Journal of Teacher Education, 41*(4), 3-11.

Perks, P.A.: 1997, *Lesson Planning In Mathematics: A study of PGCE students*. Unpublished Ph.D. Thesis. University of Birmingham.

Perks, P. and Prestage, S.: 1994, 'Planning for learning.' In B. Jaworski and A. Watson (eds.), *Mentoring in Mathematics Teaching* (pp. 65-82). London: Falmer Press.

Prestage, S.: 1996, 'Teachers' perceptions of sequencing and progression in the National Curriculum.' In D. C. Johnson, and A. Millett (eds.), *Implementing the Mathematics National Curriculum: Policy, politics and practice* (pp. 75-98). London: Paul Chapman.

Prestage, S.: 1999, *An Exploration into Teachers' Subject Knowledge*. Unpublished Ph.D. thesis. King's College, University of London.

Prestage, S. and Perks, P.: 1999a, 'Editorial: Extracting the generality.' *Mathematics Education Review, 11*, 52-55

Prestage, S. and Perks, P.: 1999b, 'Questions about subject knowledge: Learner-knowledge and teacher-knowledge.' In E. Bills and A. Harries (eds.) *Proceedings from the 4th British Congress of Mathematics Education*, July 1999 (pp. 135-140). University College of Northampton.

Prestage, S. and Perks, P.: 1999c, 'Towards a pedagogy of mathematics initial teacher education.' In E. Bills (ed.) *Proceedings of the British Society for Research into Learning Mathematics* Conference held at the University of Warwick (pp. 91-96). Coventry: University of Warwick.

Shulman, L. S.: 1986, 'Those who understand: Knowledge growth in teaching.' *Educational Researcher, 15*(2), 4-14.

Tamir, P.: 1988, 'Subject matter and related pedagogical content knowledge in teacher education.' *Teaching and Teacher Education, 4*(2), 99-110.

Wilson, S. M., Shulman, L. S. and Richert, A. E.: 1987, '150 different ways of knowing: representations of knowledge in teaching.' In J. Calderhead (ed.), *Exploring Teacher's Thinking* (pp. 104-124), London: Cassell Education.

7 MATHEMATICS STUDENT TEACHERS' RESPONSES TO INFLUENCES AND BELIEFS

Jim D. N. Smith

Sheffield Hallam University

This article forms part of an ongoing study of student teachers of secondary mathematics. The aspect reported on in this article is an analysis of the effects of the influences brought to bear upon four individual student teachers of secondary mathematics as they progress through a one-year postgraduate course of teacher training (PGCE) based at a British University. The students have differing initial beliefs about teaching, learning and mathematics. As anticipated in the literature, the student's initial beliefs survive virtually intact throughout the year. However, the study suggests that the link between initial beliefs and teaching approach is not deterministic. The study suggests ways of encouraging student teachers to employ a range of pupil activities in their teaching.

INTERNATIONAL BACKGROUND

Internationally, since the work of Lortie (1975), evidence has accumulated to indicate that teacher training is a minimal impact enterprise. Some of this evidence was reviewed in a meta-analysis of 40 previous studies by Kagan (1992). The lack of impact of training has been generally attributed to the weight of previous experience of education (as a learner) that each student teacher brings with them to their course (e.g., Powell, 1992). This is thought to lead to a predisposition to teach in particular ways and to entrenched beliefs about the nature of teaching and learning. Some have felt that these initial beliefs need to be re-examined and perhaps modified if student teachers are to make progress:

> The task of modifying long-held, deeply rooted conceptions of mathematics and its teaching in the short period of a course in methods of teaching remains a major problem for mathematics teacher education. (Thompson, 1992, p.135).

The general findings about the persistence of beliefs now have some recent exceptions. In a study of 162 secondary student teachers following a one-year PGCE, Bramald, Hardman and Leat claim that:

> The findings of the study argue that conclusions about the effects of pre-service course on student teachers' thinking are too pessimistic and need some refinement. (1995, p.30)

In the USA, Graber documents nine features that appear to explain the effects of a teacher education program that had been documented as having a high impact on the beliefs of trainee Physical Education student teachers, (Graber, 1996). Also in the USA, Joram and Gabriele (1998) concluded that making use of images and metaphors to work on student teacher beliefs made significant impacts during training. In Australia, Nettle (1998) found evidence of both stability and change in the beliefs of primary student teachers. Nettle was critical of the meta-analysis carried out by Kagan (1992.) which claimed to show a lack of impact of teacher education on student teacher beliefs.

Teacher beliefs and preconceptions are not the only mechanisms at work in influencing the student teacher's classroom approaches. Other factors include the influence of pupils through management considerations and of staff through the staff room culture (e.g., Brown and McIntyre, 1993: Gregg, 1995). The socialisation of student teachers is merely the beginning of an ongoing process, continuing through the initial training and extending throughout the teaching career (Lacey, 1977).

NATIONAL BACKGROUND

In the UK, Mathematics has been described for many years as a difficult subject to teach and to learn (for example, Cockcroft, 1982, para. 228). Many adults, having learned mathematics through traditional approaches at school, have come to view the subject with a lack of confidence and anxiety (e.g., Buxton, 1981).

In 1982, *Mathematics Counts*, a government inquiry into the teaching of mathematics in the UK was published and met with wide support. The report contained what was to become a very well known and often quoted paragraph:

Mathematics at all levels should include opportunities for

- exposition by the teacher;

- discussion between teacher and pupils and between pupils themselves;

- appropriate practical work;

- consolidation and practice of fundamental skills and routines;

- problem solving, including the application of mathematics to everyday situations;

- investigational work.

In setting out this list, we are aware that we are not saying anything which has not been said many times and over many years... (Cockcroft, 1982, para. 243)

Despite this advice, not much may have changed in some UK schools between 1982 and 1993. For example, in 1993 the UK Office for Standards in Education (Ofsted) reported that most of the mathematics teaching they saw involved pupils listening to the teacher and then working through exercises and that many pupils still experienced a very limited range of learning activities overall, (Ofsted, 1993).

It can be argued that Ofsted's comments are not research evidence, that the presence of inspectors in a classroom may modify teacher behaviour tending to make the teachers adopt a more conservative approach. However, if this is the case, the very adoption of more conservative approaches suggests that teachers expect to have these 'established techniques' valued by Ofsted. This could imply that the educational culture of mathematics teachers might not value a wide variety of classroom approaches.

Some teachers found that Cockcroft 243 provided a framework and legitimisation of their wish to include a wide variety of approaches into their classroom practice. Whilst there was no guarantee that this would be an effective strategy for teaching, it could be seen, at the very least, to lead to critical success in the 1990s;

> The most successful teachers were those who used a variety of teaching strategies depending on both the nature of the class and the learning objectives of particular lessons. More teachers had extended their repertoire of teaching styles. (Ofsted, 1993, p. 3)

> In weak schools, the range of teaching approaches is often narrow, such as where there is much exposition by the teacher, but very little opportunity for pupils to participate and respond, or where pupils are expected to learn too much on their own. In many weaker lessons, pupils spend long periods following set routines but do not learn when and how to use them. (Ofsted, 1995, para. 19)

To some extent, such observations by Ofsted are value judgements and are based upon the assumption that an extended repertoire of teaching approaches will lead to improvements in pupils' learning. At present there appears to be little research evidence to support such an assumption in a direct way. However, there is at least some evidence that an increased repertoire for student teachers does not reduce pupil performance on conventional testing, whilst it does lead to improved pupil attitudes to the subject, (Simon and Schifter, 1993).

The categories of exposition, discussion, practical work, practice, problem solving and investigational work used in Cockcroft 243 are further developed in subsequent paragraphs of that report. The categories have been taken up by the UK mathematics teaching community and are frequently used to describe teaching approaches.

For the purposes of this study the term 'teaching style' has been developed from this starting point of six approaches and has been conceptualised as the range and balance of planned mathematical learning experiences provided by the teacher for the pupils. This conception of teaching style is related to the concept of 'pedagogic content knowledge' developed by Shulman (1986). This is the distinct type of knowledge needed by teachers to transform subject content through a range of techniques, suitable examples, analogies, etc. to make content more accessible and memorable to the learner. However,

> .... as interpreted in at least most research studies published, pedagogic content knowledge would seem to imply the use of transmission methods of teaching... (Brown and McIntyre, 1993, p. 8)

The intention here is to think of a transmission mode of teaching as only part of an extended repertoire of teaching approaches. Pedagogic content knowledge is relevant, provided it is knowledge about when and how to use an approach selected from an extended repertoire.

The motivation behind this study is an attempt to extend the repertoire of pupil learning activities in use by student teachers of secondary mathematics and increase the student teachers' options for professional choice. However the initial teacher training course may be designed to modify the cycle of teacher replication, it must be recognised as a difficult task in the time available on a one-year PGCE programme.

THE UNIVERSITY COURSE

The University works in partnership with over 100 local schools to provide a 36-week academic year teacher education programme that is two-thirds school-based and one-third University-based. During the study, the pattern of school time and University time over the 36 weeks was as indicated in Table 1.

Weeks	Monday	Tuesday	Wednesday	Thursday	Friday
1 and 2	University	University	University	University	University
3	School	School	School	School	School
4 to 7	University	University	University	School	School
8	University	University	University	University	University
9 to 18	School	School	School	School	School
19 to 20	University	University	University	University	University
21 to 24	University	University	University	School	School
25 to 34	School	School	School	School	School
35	University	University	University	University	University
36	School	School	University	University	University

Table 1: The University PGCE Course

Weeks 1 to 18 were associated with placement school 1, and include serial days in Weeks 3 to 8 leading in to a block practice in Weeks 9 to 18. Weeks 19 to 36 were associated with placement school 2.

When encouraged by the University tutors to adopt a variety of teaching approaches in school mathematics lessons, many student mathematics teachers have experienced a clash of culture and beliefs about the nature of teaching and of mathematics. They have found themselves trying to respond to conflicting demands from their host schools and the University. This experience of a clash of culture was confirmed in an earlier study (Smith, D. N., 1996). In that study the particular cohort of one-year PGCE mathematics student teachers tended to side with the University viewpoint and were critical of the narrow range of teaching approaches adopted by the mathematics teachers in many of their placement schools.

Previous stages of this research programme (Smith, D. N., 1996; 2001) indicated a number of pressures being brought to bear upon student teachers' choice of teaching materials, pupil activities and learning experiences. In summary, these pressures arose from the following sources:

1. University-based tutors.

2. School-based class teachers and mentors.

3. Pupils (indirectly, through management considerations).

4. The student teacher's view of mathematics.

5. The student teacher's views and experiences of teaching.

6. The student teacher's views on the nature of learning.

Alongside these 'traditional' pressures, in 1998/99 there came the introduction of the UK's National Curriculum for Initial Teacher Training and the associated competency-based documentation (Department for Education and Employment, 1998). These developments also sought, in part and indirectly, to put pressure onto student teacher's choice of classroom approaches.

In an effort to understand the school viewpoint more clearly, part of the previous research (Smith, 2001) was focused on analysing advice offered to student teachers by school staff. This clearly indicated that the advice offered to a cohort of PGCE student teachers of mathematics by school staff was predominantly concerned with class management, rather than considering alternative teaching approaches. Where alternatives were offered, these were narrowly focused on improving expositions, examples and exercises rather than on widening the repertoire of teaching approaches. Apparently, the tradition of teaching predominantly through exposition, examples and exercises was being perpetuated.

At this point it was felt useful to attempt to discover in detail how individual student teachers of mathematics responded to conflicting demands made of them, and how they resolved these in practice through their choice of pupil activities.

Perhaps more intensive, more focused studies of individuals will help us to understand the concerns of novice mathematics teachers and the roles these concerns play in the process of becoming a mathematics teacher. (Brown and Borko, 1992, p. 230)

The following sections describe the aims, methodology and results of a study designed to explore the influences on student teachers' choice of pupil tasks.

AIMS

The main aim of the study described in this paper was to examine in detail some of the mechanisms influencing individual secondary mathematics student teachers' selection of pupil activities. In particular, the focus was on the ways in which student teachers devise their own pupil activities, adapt existing activities, make use of pupils' texts and guidance offered by class teachers, mentors and University tutors. The intention was to add to research in the area by devising some additional strategies to broaden student teachers' repertoires of classroom approaches.

METHODS AND DATA COLLECTION

Through the process of carrying out this research, it has become clearer that learning does not take place in isolation, but is influenced by a community. For example, in the case of student teachers the immediate community includes mentors, tutors, pupils, peers, friends and family. The theoretical framework used in the research reported in this paper has been strongly influenced by views of learning as socialisation (Lacey, 1977, p.13), learning as enculturation (Calderhead and Shorrock, 1997, p.11), and learning as participation in a community of practice (Lave and Wenger, 1991). The choice of methods has therefore involved trying to obtain data from a variety of sources within the community of the student teachers' practice. This theoretical ideal position has been balanced against the time and opportunity constraints and resulted in a selection of approaches, which were intended to be both achievable and produce a variety of perspectives.

In order to manage resource considerations so as optimise opportunities to work in depth and to gain a useful variety of evidence, the sample size had to be relatively small. The sample was chosen in order to illuminate similarities and differences in the training experiences. A promising dimension was that of absolutist / fallibilist views of mathematics (Ernest, 1991) which might be associated with an instrumental or relational dimension of mathematical understanding (Skemp, 1971) and teaching of the subject matter (Lerman, 1983). However, this dimension was not expected to form a substantive background to the research; its use was merely a device intended to help facilitate a variety of viewpoints from a relatively small sample of student teachers.

Reassurances were given to the entire cohort concerning ethical issues and confidentiality of the research. An *Initial Beliefs Questionnaire* was devised and administered to the entire cohort to identify their pre-training views on the nature of mathematics, teaching and learning. This would provide a benchmark for later comparisons when looking for changes over the course.

The *Initial Beliefs Questionnaire* was used to identify four students with diverse initial views, who were invited to take part in the research work. These are referred to as Emma, Quentin, Keith and Tony. The actions shown in Table 2 were undertaken to collect appropriate data using a variety of techniques, each designed to focus on an aspect of the research interest. Hence a variety of approaches was used; questionnaire, semi-structured interview, documentary analysis and lesson observation to try to improve internal validity. To incorporate triangulation the views of other observers, i.e. class teachers and mentors, were collected and analysed.

The observations are presented in approximate chronological order to allow the reader to sense the developments made by individual students over the course whilst facilitating comparisons between the individuals at different points in the year.

Question/focus	Methodology
Analysis of student teachers' selection of pupil tasks.	Semi-structured interviews to examine the source of activities used in lessons on block practice.
How does experience in the classroom affect the choice of pupil activities?	Use and documentary analysis of reflective diaries and lesson self-evaluations. Analysis of critical incidents.
What advice is being offered to the student teachers?	Documentary analysis of written advice from teachers. Ongoing interviews concerning changes in views or practices. Teacher questionnaire/ student questionnaire.
What is the relationship between an individual student teacher's discourse about teaching and their classroom practice?	Lesson observations. These were also helpful in building an image of the student teacher's class teaching persona, in developing a working relationship and in sharing an understanding of terms such as 'pupil discussion', 'practical work', etc.
Are there changes in beliefs over the year?	Repeat *Initial Beliefs Questionnaire* at the end of the year and compare results.
How do student teachers perceive the issues around teaching style?	Ask students to comment on three math's teachers they have seen in action, looking for differences and similarities. Analyse the discourse for students' own descriptors of teaching style.

Table 2: Research questions and methodology

THE STUDENTS

These four students have a variety of backgrounds, different expectations, different needs and differing concerns about learning to teach. As Haggarty (1995) notes in her study of ten PGCE students, care must be taken to consider the effect of the nature of individual students. Each has a different personality and may well react in different ways when placed in similar contextual situations, e.g., school placement.

Emma was aged 22 at the start of the course and had moved to the UK from South Africa when aged 14. She completed her final two years of secondary schooling in a local comprehensive, then took A Levels before obtaining a Degree in Electrical Engineering.

121

Quentin was aged 50 at start of the course, having a working background in marine and mechanical engineering before becoming a self employed retailer.

Keith was aged 30 at the start of the year and had been schooled locally before obtaining a B.Sc. Mathematics. He was then employed for eight years as manager of chain of licensed betting shops. In his spare time, he became an Assistant Scout Leader.

Tony was aged 22 at start of course and had been schooled locally before coming to the University direct from taking his B.Sc. Mathematics elsewhere. He had intended to teach since completing his A levels, and had undertaken some Education options in his degree route.

It might be helpful to consider Emma and Tony as being 'traditional' teacher training students, following Powell's (1992, p. 227) classification, in that they were undertaking a postgraduate teaching qualification immediately following their undergraduate courses. Quentin and Keith would be classed as 'non-traditional' by Powell, who illustrates different effects of initial teacher education on these two categories of student. Powell suggests that the traditional students, Emma and Tony in this case, would be more influenced by their own schooling. Powell also suggests that the non-traditional students, Quentin and Keith in this case, would be less influenced by their own schooling and more likely to be influenced by a body of professional pedagogical knowledge.

An overview of the data from this study is given in the following sections. Some data is presented in detail so as to allow comparisons and other forms of analysis to be made.

INITIAL BELIEFS (WEEK 1)

This stage of data collection involved in-depth semi-structured interviews to determine students' initial views of teaching approaches. During the interviews, field notes were taken and afterwards these were typed into a draft statement describing each student teacher's initial views. Respondent validation of this statement was obtained and disagreements were resolved by redrafting until the student teacher agreed that an accurate description had been developed. These statements are shown below, after modification to put them into the past tense. The early use of technical language by the student teachers reflects its appearance in the initial questionnaire, in discussions held during the normal lecture programme and in the interviews.

Emma believed that mathematics is discovered in the world and that it is present all around us. She thought that the learner therefore had to acquire pre-existing facts and concepts from books, their teachers or through experimentation with real objects. She thought that lessons based on a 'discovery' model would be most helpful to pupils. She expressed the view that the teacher should both provide explanations and learning opportunities for pupils, and that pupil discussion is essential to learning.

Emma believed that mathematics is based upon a set of reliable axioms to which logic is applied. In some cases, she accepted that the axioms might be somewhat arbitrary "especially when considering ideas such as 'i' in complex numbers and some of the conventions surrounding zero and one". Emma emphasised the procedural, claiming that "Mathematics consists of a set of laws/rules which are applied in certain situations in order to arrive at a solution."

She believed that mathematics is culturally independent in the sense that "It is like an international language", and to be value free: "Maths itself isn't intrinsically good or bad, its value lies in how you use it." Emma believed that mathematical definitions are important and that pupils should learn them. However, she also felt that pupils need to work out their own understanding of definitions and that learning should proceed towards clear definitions. "Definitions are important, but pupils need to get their own ideas about them."

Her initial beliefs about teaching were identifiable as a transmission orientation: "The transfer of knowledge and the enthusiasm for a subject from teacher to pupil."

Quentin held an image of mathematics as providing a collection of mental tools: "I look upon mathematics as the tools for other subjects." This may have reflected a utilitarian view of mathematics that had its roots in Quentin's engineering background. It appeared to lead Quentin to a model of teaching that was based on introducing pupils to tools of increasing sophistication, power and complexity. In his model, pupils learn to handle an ever greater number of these mental tools and develop an increasing ability to use each correctly and purposefully.

In defining mathematics Quentin referred to it as "the science of expressing and manipulating numerical information" and discussed sciences as having "order, history, rules and technology." He believed that mathematics is discovered in the world around us, in a similar way to other sciences, and agreed with the statement that mathematics and science are studies of pre-existing phenomena. There were aspects of Quentin's thinking at this early stage in the year which suggested that he would be inclined to use some 'learning by discovery' in his teaching approaches.

Quentin's general view of teaching emphasised that both the transmission and reception of knowledge and experiences are important.

Keith adopted an absolute view of mathematics, claiming that "maths is based on applying logic to come up with an answer." He believed that mathematics is based on reliable axioms rather than convenient fictions and is the study of truth and a search for absolute truths.

He also viewed mathematics as being discovered, rather than created and that mathematics is all around us in the world: "A large proportion of maths has already been discovered by mathematicians and is out there to be discovered." Keith believed that mathematics exists in books, computers and in teachers' knowledge. This led Keith to believe that pupils should often be using discovery and investigative approaches to learn the subject. The precise methods should vary, he thought, with the mathematical content of the session.

Keith viewed mathematics as hierarchical and believed that this has implications for teaching; "You need to grasp the basics to build mathematics up."

Tony had a strong belief in teaching as being a "passage of knowledge and experience from an individual to others". He referred to a metaphor of a teapot used to pour tea into a number of cups to model the teaching process. However, he felt that it was "up to individual pupils to show that they could do more than simply receive knowledge".

He believed that mathematics is founded upon reliable axioms, and that logic is used to create, sustain and prove mathematics in relation to these true axioms. Definitions were important to Tony, and he believed that "Definitions are quite useful. When you go further in maths, they are unchanged. They are concrete, I like things which don't change."

He believed that mathematics is discovered rather than created, as it is all around us in the world. However, Tony was prepared to concede in discussion that some degree of idealisation might be used in generating mathematical concepts (e.g., a rectangle).

He believed that learners have to construct their own understanding of the subject: "People learn in different ways, there is no point in forcing someone to learn in a particular way." At the same time, he believed that teachers should explain mathematics to their classes rather than provide learning opportunities. He commented that this was a personal preference related to his own learning style "I like things to be clearly explained to me ...." Tony felt that there were some contradictions inherent in holding both of these views simultaneously and thought that he might need to employ a wider repertoire of teaching approaches than direct explanation.

EARLY EXPERIENCES (WEEKS 1 TO 8)

Emma

In her first attempt at planning a lesson, Emma researched a range of five textbooks for ideas and discussed with the class teacher. The lesson introduction began with the opening remark: "Percentages are all around us in the world". This related closely to Emma's beliefs about mathematics. The first pupil activity was to copy notes from the board as to how to convert between fractions and percentages. This was presented as a procedure to be learned as were several other procedures for working with percentages and related closely to Emma's 'transmission of procedures' model of mathematics teaching.

Quentin

Two issues seem important to Quentin at this stage of the year, class management and his own subject knowledge.

Journal extracts from his first day in school show a focus on pupil behaviour and class management, for example "Y11 Set 5 – very disruptive class – low content of lesson. Jamie threw books around the room ..." and "Y8 Set 2 class – well behaved – working on shape ..." He also summarised advice from his mentor and the Head of Mathematics about class management: "Control is more important than maths. Without control maths cannot be taught."

Some worries surfaced about Quentin's own knowledge of the subject when he spent a lesson with top set Year 11 pupils (students aged 15-16). He wrote, "I need more exposure to maths at this level before I would be confident to take a class unless it was a very well prepared specific topic".

Keith

During the early stages of the course, Keith did not appear to have many concerns. In an interview during Week 7, I asked him about the advice that he had received from class teachers about the choice of pupil activities, resources and teaching approaches. He summarised this as:

Always have backup material for extension work available.

Collect activity resources in straight after the activity.

Check resource list for availability.

Keith added that he had not been advised on a range of alternative teaching approaches.

Tony

Tony appeared to have serious concerns that were not clearly focused at the start of the year, writing in his journal during Week 4 "I am very worried about being able to deal with the whole experience and expectations."

TEACHING PRACTICE (WEEKS 9 TO 18)

During this stage of the course, students were working full time in their placement schools. Data was collected from lesson observation summaries written by mentors, from videotaped lesson observations, student's journals and teaching practice files. The selection of data presented here has been made based on apparent relevance to the student teacher's use and choice of pupil activities.

Emma

In Week 18, I conducted a lesson observation. Emma was working with a Year 7 (students aged 11-12) mixed ability class on the area of triangles. Two main activities were planned for the session. The first of these making use of pupil text books on area (to fit in with departmental scheme of work), the second being a number game (chosen for reasons of creating variety and sustaining motivation). The area booklets were incorporated into a standard approach with examples, explanation and exercises.

The transition to the second activity was muddled. The second activity was number work using a *Countdown* game adapted by the student teacher from the eponymous TV programme. A lot of pupil interest was produced here, but the management and sustaining of this was somewhat haphazard. This was a potentially very motivating activity but it needed further structure to cater for differentiation and involvement of all pupils.

Quentin

During Week 13, I undertook a lesson observation, which was video recorded for later analysis and reflection. The class were in Year 8 (students aged 12-13), and were Set 2 out of 8. Quentin's planned learning outcomes were "Recognise a pattern in the number table – understand that a formula can be derived to predict the number of squares/colours for any pattern 'p'." The main pupil activity was called *Traffic Lights*. Quentin had found this in the departmental reference books, adapting the task by producing his own worksheet to accompany it. At the time Quentin stated that "it was a special lesson but not untypical." Good dialogue was created with the class, the lesson was well structured but there were some problems with timing. Quentin said that in his attempt to fit the lesson into 55 minutes he rushed the work too quickly.

None of the lesson observation reports made by class teachers or mentors commented on the choice of pupil activity. The comments were largely focused on class management issues and clarity of exposition.

Keith

During Week 18, I undertook a lesson observation that was video recorded for later analysis and reflection. The class were the top set in Year 7 (students aged 11-12). The activity was an investigation into *Match Stick Patterns*, taken from the departmental scheme of work, and adapted by Keith with his own presentation using an OHP. This is an activity that had been discussed in University sessions (adapted from Smith, J., 1996). The pupils worked very well on this after a very brief introduction from the student teacher.

In addition, during Week 18 an observation was carried out of Keith teaching a Year 10 class (students aged 14-15). The pupils were learning about perimeters of rectangles and composite rectangle. This was a more conventional exposition, examples and exercises lesson.

Tony

During Week 11, an observation was carried out by the class teacher of Tony working with a Year 8 class (students aged 12-13). The topic was probability, and Tony used an activity introduced at University, "Table Races" (Smith, J. 1996). The class teacher wrote: "The lesson was a well -planned piece of practical work, intended to develop pupils understanding of probability. The grids to collect the data were clearly illustrated and explained, as was the procedure for the experiment."

In Week 15, I observed Tony working with a Year 7 mixed ability class (students aged 11-12) on equivalent fractions. After a discussion with the whole class about their previous work on finding fractions of whole quantities (e.g. two thirds of 27 sweets), Tony used the 'shading' technique to illustrate fractions and their equivalence. The question and answer dialogue of the introduction led into an exercise from the pupil text.

During Week 16, Tony taught a Year 10 set (students aged 14-15) to draw graphs using graphical calculators. The lesson was observed by the class teacher who wrote: "With a difficult group such as this, the lesson was definitely risky with the need for practical involvement and the possibility of equipment failure, but it was broadly successful. Generally a good try with a difficult topic and definitely worth the risk. Well done."

SUMMATIVE REPORTS (WEEK 17)

These are reports written by mentors on the progress of the student teacher over the first teaching practice block. For each student I have included the comments that appear to directly refer to pupil learning activity choices.

Emma

Emma has demonstrated that she can teach by exposition and encourage individuals to explore concepts.

Quentin

Variety of teaching styles observed (exposition, individual teaching, teaching small groups).

Keith

Keith's planning is quite thorough but he doesn't always identify appropriate transition points, nor does he plan sufficiently for differentiation. His explanations are clear. He needs to be more selective when setting exercises to do, to allow for the different abilities within the group. [*There were no other comments on pupil activities.*]

Tony

Produces sound plans addressing target outcomes, resources, activities and motivation found within the groups.

[Tony is] aiming to broaden the range of strategies used.

[Tony is] becoming more confident with difficult group - willing to try 'risky' strategies. [*In this extract, the word 'risky' appeared to be closer to a speculative gamble rather than to injudicious.*]

INTERVIEWS REGARDING SEMESTER 1 TEACHING (WEEK 19-20)

These interviews were carried out with the student using their teaching practice file to provide a stimulated recall of their planning decisions. Since it was not practical to examine all lesson-planning decisions, a sample was made by subjecting the lessons from Weeks 10 and 17 to analysis. These weeks were chosen as they were in the middle of teaching blocks and it was expected that routines might have been established. Other lessons were also discussed, but less rigorously analysed. The interviews were summarised, with a focus on the choice of pupil activities.

Emma

There was a pervading similarity in the structure of Emma's lessons, with a few minor variations. This was characterised, by mutual agreement, as follows:

• Review of previous lesson in the form of question and answer dialogue with the whole class, except when beginning a new topic.

• Further dialogue with the whole class on new material. Sometimes this would be based around a small-scale activity such as using a number line to represent probabilities, or getting a pupil to face north and rotate to other compass directions to represent angles.

• Examples, explanations and sometimes notes.

• Exercises from commercially produced sources or own worksheets

• Sometimes a small-scale activity at the end of the session, typically on numeracy, motivated mainly by wishing to pattern the lesson to sustain pupil interest.

The activities from the sample lessons were recalled to have come from the following sources:

Habitual review	29%
Own idea	25%
Pupil text	21%
Host Department	14%
University tutors	7%
University peers	4%

If one takes the 'habitual review' as being Emma's own idea, then she can be seen to be developing most of her teaching activities herself, but also drawing on a range of other sources.

Constraints on the choice of pupil activities were recalled to be:

• Due to rooming with Year 7, where an old and oddly shaped classroom limited opportunities for all pupils to see the board, limited presentation space at the front and provided other physical constraints on the teaching.

• Serious concerns about pupil management with a Year 10 class led to hesitation about choosing unusual activities and a tendency to play safe with textbook based materials. However, Emma did manage to lead a computer - based session on data handling with spreadsheets.

Quentin

Quentin's lessons did not appear to fall into a regular pattern. He tried to respond to the particular needs of the different groups as expressed by the class teachers and mentor and his own judgement of the pupils' ability and co-operation.

For Year 8 Set 2/8, Quentin claimed that "a very high proportion of the lessons were planned around the text book, and this was supplemented when difficulties were found." The class were reasonably amenable and worked well with Quentin. "I felt that I was really able to teach this group." There were constraints due to the departmental use of a textbook programme. "I had three weeks to finish the book before it was needed by another class."

For Year 10, Set 5/8, there was little guidance from the school on precisely what to cover; the scheme of work given amounted to a side of A4 rough notes from the class teacher. The pupil textbook was used at times for source material, and frequent reference was made to the class teacher to try to get the level of difficulty right. Quentin found initially that he pitched the work at too high a level.

For Year 7 mixed ability: The department used the Oxford Mathematics scheme, with whole classes split into working on the 'long book' or the 'short book' as a strategy for managing the mixed ability range. Despite the text books being well presented and containing varied teaching approaches, Quentin found this to be very difficult to manage as sometimes it meant addressing part of the group while the rest were expected to continue with their work. Consequently, Quentin sometimes supplemented the bookwork with whole class work on supplementary material that he designed himself.

The activities from the sample lessons were recalled to have come from the following sources:

Pupil text	44%
Own idea	44%
Host department	6%
Reference text	6%

Quentin appeared to be relatively proficient in developing his own ideas, but these were often supplementing the bookwork rather than being independently initiated. However, he did add "I have used several activities with my classes which were not in the schemes of work. Some activities which come to mind are: number patterns from triangles using match sticks, pupil rectangle [see *Smith, J., 1996*], a pendulum activity timing 20 swings for different length strings."

Quentin commented that he liked many of the pupil activities shown in the college sessions. However, he felt unclear how they might fit into a departmental scheme of work and how he might justify the place of these in his teaching.

Keith

Keith found the departmental scheme of work "helpful, straightforward and easy to follow." There was plenty of encouragement from his mentor, who was also Head of Mathematics, to experiment and try out ideas.

There was no clear pattern developing in the learning-activities adopted by Keith other than that of variety. From the sample of lessons, the sources of pupil activity were reported as follows:

Own idea	50%
Pupil text	22%
Host department	28%

During these lessons, a good range of activities and tasks was offered to pupils. These included explanations, dialogue with the teacher, a Multilink task to represent fractions, pupil text booklets, an investigation into colouring badges, a practical task to measure circle circumferences, using a software package to reinforce number bonds, worksheets produced by the student, use made of OHP and board, practical task estimating coins in a stack, and routine practice from pupil texts.

Tony

Tony claimed that he received little help with planning the lessons at this school, because it was assumed that he would ask for help when he needed it. Whilst willing to ask for help Tony did not feel that he needed any and so did not receive much assistance with planning.

On a straightforward count of activities the following sources of pupil activities were identified with these frequencies:

Own ideas	76%
Pupil text	18%
Host department	6%

Tony claims to have worked out the vast majority of his teaching ideas himself, often using the pupil text as a source of ideas. He showed a developing level of pedagogic content knowledge. For example, he took an idea for using probability spinners from a class textbook and demonstrated to the class how to construct a spinner and then how to modify it to introduce bias. The pupils then experimentally investigated the bias that they had predicted to exist in their own spinners.

There were some interesting and relevant remarks made by Tony about planning lessons for "a stubborn and difficult [*Year 10*] class." With this group, he tried a wide variety of approaches over the practice, hoping to find a way to provoke the pupils' interest and motivation. This included planning the lesson employing graphic calculators to explore straight-line graphs. The lesson required a substantial amount of preparation and careful design of worksheets, visual aids and clear instructions.

There was encouragement from the school staff to try this, but little guidance with the detail of the planning. In the event, this worked relatively well. In another lesson, Tony devised a worksheet on area and perimeter which provided a very practical context, again with the intention of raising interest and thereby motivation.

With a bottom set Y9 class Tony claimed to have been advised by the class teacher to use a very didactic and direct instruction-based approach with the suggestion that anything else would not work. However, Tony did ask pupils to carry out some structured investigations into angle properties with sufficiently encouraging results to try out further exploratory work with the pupils. This is clearly an interesting development from the point of view of the aims of this study and so I asked why he had done this kind of work in the face of advice not to do so. Tony claimed that:

> It was actually Paul [*the class teacher*] who said that spoon feeding the pupils was the way to go. I was met by [*ironic*] comments like "Good luck!" when I mentioned the idea of investigating. In some way, this inspired me to go ahead. I wanted to succeed in some way to show Paul that it was possible, one just for self satisfaction and two so that he might try in future.

Tony linked this to motivating the pupils through making the work more interesting. He reported that pupils gained much satisfaction from noting for themselves that the angles on a straight line added to 180° every time.

INTERVIEWS DURING SEMESTER 2 AT THE SECOND PLACEMENT SCHOOL (WEEKS 25 TO 34)

These were carried out at various times during the second teaching practice. Each interview involved one or more lesson observations. Students were invited to discuss teaching approaches in their own words, by reflecting on teaching that they had observed, then comparing and contrasting three teacher's ways of working in the classroom. The student's lesson evaluations and planning were examined by a stimulated recall of the sources of pupil activities used in planning lessons in Weeks 26 and 30.

Emma (Week 31)

When asked to compare and contrast three teachers' ways of working in the classroom, Emma mentioned the status of each teacher first (i.e. Head of Department, Deputy Head Teacher, class teacher). She claimed that "all three teachers try to make the work interesting...Two of the teachers use a quick introduction and get pupils working quickly ... one uses humour to encourage pupils." Emma commented: "It's difficult to describe the differences; they have different ways of seating pupils, in pairs or fours or a mixture of the two. They spend different amounts of time on the numeracy strand, [*one teacher*] lines pupils up outside the door, another uses the OHP more... two of them use more practical activities."

Emma had adapted lesson-planning guidelines provided by the second placement school, based in turn upon advice from Ofsted. She had to incorporate a departmental requirement of a 10 minutes numeracy session into each lesson. When asked to comment on the style of lesson planning that she developed over the year, Emma wrote: "Due to an existing policy in the Maths Department my lesson planning changed to 1. Reception and registration, 2. Mental Maths spot, 3. Introduction, 4. Main activity/ies, 5. Conclusion, 6. Dismissal. I believe in time I shall develop my own style." Her lesson evaluations were extensive.

In this placement, Emma had continued to use a variety of pupil activities. Analysis of the sources of ideas for pupil - activities in the sample weeks 26 and 30 showed:

Own Idea	27%
Host Department	44%
Pupil Text	17%
University tutors	11%

Most of the ideas for Emma's pupil activities came from the host department's scheme of work. These included activities involving Möbius bands, plastic cubes, Logo for Multilink, *Skeleton Cubes*, testing for fairness in games, an Internet session, Braille, and dominoes.

Emma made use of whole-class discussion, pair work and group work, so is aware of the range of possibilities. (Lesson Comments from class teacher).

Emma had persisted with active learning approaches with her Year 8 Set 3 out of 4 despite having some class management difficulties: "The main reason being to try to improve the pupils level of motivation." For example, in Week 30 for a lesson on angle properties of triangles Emma chose a practical task of cutting up paper triangle and rearranging corners. "It didn't go well due to the time of day and some individual pupils. The problems arose before the practical work, but I continued with it any way." The fact that the host department was committed (at least in its documentation) to use a range of teaching approaches may have been helpful in encouraging Emma to use them.

Quentin (Week 32)

When asked to compare and contrast three teacher's ways of working in the classroom, Quentin talked about the level of organisation of the classes. He commented on how closely or otherwise the teachers followed the departmental scheme of work. He described one teacher as being "more old fashioned" and "making heavy use of examples and exercises." Another teacher was described as giving a very clear and precise presentation of the lesson content, which was likely to be the same "no matter who was sat there." Quentin claimed that the teacher made little or no attempt to elicit pupil responses or to use these in the development of the exposition.

The analysis of sources of activities during Weeks 26 and 30 revealed:

Host department	60%

Pupil text	30%
Own ideas	10%

Quentin seemed quite reluctant to use his own ideas or those suggested in University sessions. My field notes recorded at the time include the observation, agreed by Quentin that "what seems to help Quentin most is having a range of activities built in to the scheme of work."

Keith (Week 31)

When asked to compare and contrast three teachers' ways of working in the classroom Keith made the following observations. Teacher 1 was described as differing by his mannerisms, voice, and high level of enthusiasm, use of questioning and glaring eyes. Teacher 2 differed in the use of "... more games and activities to get kids interested and motivated and he sits down more often at his desk." Teacher 3 had a quieter voice, and pupils responded well to her. She used more exposition than the other two, and then circulated round the room to help pupils.

Activity sources Weeks 26 and 30:

Pupil text	27%
Own idea	27%
Host department	27%
University tutors	18%

This suggested that Keith was drawing on a range of sources, with no particular reliance on any one source. There were a number of activities used over the practice, including using Multilink in Shape and Space work and in investigating the volume of cuboids with Year 7; a *Think of a Number* introduction to equations with Year 7 which "went really well"; a practical task measuring circles with Year 8, a practical algebra task *Matchsticks* with Year 8; the use of weighing scales to demonstrate balancing equations with Year 8; the use of TV schedules, bus and train timetables for a lower set Year 9 class working on the 24 hour clock; Logo with Year 7 and Year 10; probability *Table Races* with Year 9.

The lesson observation report forms from mentor and class teacher, five in all, were examined for comments on the choice of pupil activities. There was a strong focus on class management, teacher exposition and many comments on the lack of differentiation in the chosen pupil exercises or worksheets. There was no discussion of alternative teaching approaches or pupil activities.

In the summative report Keith's mentor wrote:

> An extended repertoire. Has shown some sound skills and has improved throughout the practice. His relationship with the pupils has been very good; with experience he will be able to develop and use a wider range of resources and methods to enhance learning.

Tony (Week 27)

In Week 27 I had a meeting with Tony, who came in to talk to me about leaving the course. My notes at the time included: "The basic problem appears to be that he is not enjoying the teaching, and says that part way through lessons he often wishes he was somewhere else." In his first school placement, Tony put this down to the fact that pupils were not very co-operative at times. At the second placement school the feeling of being "in the wrong place" has persisted, despite the pupils being more amenable. He feels that he would be willing to continue to do the work if he was getting more job satisfaction from it." Despite not having a job to go to, Tony left the course at this point. Perhaps this was another indication of the willingness and ability to take risks that had been evident in Tony's teaching.

STUDENT'S REFLECTIONS AT THE END OF THE COURSE (WEEK 36)

Emma

In the final questionnaire, Emma expressed the view that her beliefs about mathematics, teaching and learning had not changed at all over the year. This was confirmed by comparing the final responses with the *Initial Beliefs Questionnaire*.

Emma was asked to suggest improvements to the course that might encourage students to use a broader repertoire of pupil learning-activities. She suggested "Early exposure to teaching pupils and a focus during observation on teaching styles. Seeing professional teachers using a range of styles also inspires. Also, seeing a teacher use the same teaching styles for all topics would also make the point!"

Quentin

There was very little change evident in Quentin's beliefs as evidenced by both his own comments on the subject and by a comparison of his beliefs as expressed in the initial and final questionnaires. He commented that "I still consider mathematics a collection of tools. This view has probably been reinforced over the year."

> I think my view of a teachers' role has not changed. Learning by discovery or experiment is more enjoyable and more memorable. I am more aware of the problems of control when trying to run a lesson of discovery. I would get more satisfaction from one successful activity based lesson than a string of quiet, well controlled 'conventional' lessons

Keith

Keith reflected on his initial beliefs and noted that his views had not significantly changed. This was confirmed by comparing the initial and final questionnaires.

> After observing lessons on both teaching practices most teachers tended to present the lessons by teacher exposition followed by pupil tasks. Perhaps these early lessons should show a variety of teaching styles so trainees are more aware and confident to use them once they start teaching themselves.

DISCUSSION

The student teachers used a variety of sources for their pupil activities, and were either largely reliant on their own ideas (Keith, Emma and Tony) or upon the departmental scheme of work (Quentin). Where student teachers had claimed to develop activities themselves this suggested that direct influences on their choices may have been limited. However, it may have been the case that there were influences on the type of activity, rather than the specific activity. This might provide an opportunity for further research.

Quentin and Emma seemed to have benefited from the inclusion of a range of pupil activities in their host department's scheme of work. This may be something to be encouraged within the University partnership with schools. This complements the finding that these student teachers were not often advised of alternative teaching approaches and suggests that partnership schools be encouraged to be more active in putting forward alternative strategies.

The most interesting finding for me was that each of the student teachers, for various reasons and with varying degrees of confidence, actually did employ a variety of pupil activities in their teaching.

There is a similarity here with Haggarty's finding in her study of 10 PGCE students:

> In addition, although they all learnt to use a range of styles of teaching in the classroom, this in itself did not necessarily significantly alter their views about mathematics. (1995, p.120)

Emma and Keith seem to have used an extended repertoire of teaching approaches from the perspective of wishing to raise pupil motivation. Both of these students introduced their own ideas and ideas from University sessions into the school placements. Emma seemed to prefer fitting the activities into a simple lesson format.

Tony used an extended repertoire of teaching approaches. This was despite some opposition from teaching staff, and may indeed have been partially provoked by such opposition. Tony felt convinced that an extended repertoire would be more motivating for pupils than the normal offering by the class teacher. Tony showed himself to be able and willing to take risks in his teaching career where other students were more cautious. In particular, he appeared willing to try pupil activities where the outcome was not predictable in terms of management.

Quentin appeared to value the practice of using a wide variety of activities, but was often insecure in choosing these for himself and appeared reliant upon the scheme of work to suggest or require particular pupil activities. This may have been related to his insecurities in subject knowledge. Nevertheless, he did introduce some of his own ideas to support the text-based material and some that were beyond the departmental scheme. This appeared to be when he felt confident in the role of the activities to support pupil learning, rather than simply as an aid to pupil motivation. Quentin commented after the end of the course:

At the beginning of the course we were shown activities without them being related well enough to topics, or shown where they would lead. I was not experienced enough at that stage to understand how to use them in the classroom. If we had looked at a couple of activities more rigorously before the first teaching practice I may have had more confidence to try them.

Incidentally, we have now incorporated this suggestion into course planning for future years.

None of the students explicitly related their use of teaching approaches to their beliefs about mathematics, although it is possible that they were unconsciously influenced by these beliefs. Three were motivated themselves to use a variety of pupil activities by considerations of pupil motivation. The fourth appeared to use his choice of pupil activities more in relation to the mathematical content of the lesson and was more reliant on the departmental scheme of work.

CONCLUSIONS

There are limitations to the methods used in any study and there are limitations to the space available for publication. The attempt to incorporate the views of class teachers and mentors was worthwhile, but it is possible that these might not always be the same as the views expressed verbally to the student teachers. Student teachers appeared sometimes to take on board different messages from those apparently intended. This was an ethical issue: a communication problem existed between student and mentor; I knew about it but had given guarantees of confidentiality which could not be (and were not) broken. It was also a methodological problem as it was not possible in such situations to reach a consensus view through discussion.

In contradiction to some expectations, (e.g. Bruner, 1996; Ernest, 1989; Zeichner and Liston, 1996) this study supports the view that there is not a deterministic relationship between beliefs about mathematics and the student teacher's classroom teaching style. Beliefs about mathematics do not inevitably lead to a particular teaching style or to a narrow range of pupil learning-activities.

The implication of this finding for my own University teaching is quite significant. For many years the University tutor team had felt it appropriate to begin as early as possible to challenge student teachers' beliefs and preconceptions in order to help the students to see the value of an extended repertoire of teaching approaches. It would appear that this might be inappropriate and unnecessary. It may be inappropriate because of the persistence of such beliefs despite our best efforts. It may be unnecessary because the student teachers did adopt a range of approaches despite holding different beliefs about mathematics. Some redesign of the course has now taken place to build upon concerns with class management and pupil motivation.

Future research will build upon and test the links tentatively established here that:

- Concern regarding class management leads to an interest in pupil motivation.

- Concern with raising pupil motivation leads to an increased repertoire of teaching approaches.

I feel that I have good reason to consider that this may be a more effective strategy than trying to tackle beliefs about the nature of mathematics, teaching and learning which have proved to be resilient in these individuals as well as in most research studies. It will be appropriate to reflect on what happens to student teachers' range of approaches once pupil management issues are largely resolved and are no longer a driving force promoting variety.

ACKNOWLEDGMENTS

I acknowledge with thanks the support of Peter Gates, Nottingham University, Denise Smith, the student teachers involved and the many other colleagues who expressed an interest in the study.

REFERENCES

Bramald, R., Hardman, F., and Leat, D.: 1995, 'Initial teacher trainees and their views of teaching and learning.' *Teaching and Teacher Education, 11*(1), 23-31.

Brown, C. A. and Borko, H.: 1992, 'Becoming a mathematics teacher.' In D. A. Grouws (ed.): 1992, *Handbook of Research on Mathematics Teaching and Learning* (pp. 209-239). New York: Macmillan.

Brown, S. and McIntyre, D.: 1993, *Making Sense of Teaching*. Buckingham: Open University Press.

Bruner, J. S.: 1996, *The Culture of Education*. Cambridge, MA: Harvard University Press.

Buxton, L.: 1981, *Do you Panic about Maths?* London: Heinemann.

Calderhead, J. and Shorrock, S. B.: 1997, *Understanding Teacher Education*. London: Falmer Press.

Cockcroft, W. H.: 1982, *Mathematics Counts*. London: HMSO.

Department for Education and Employment: 1998, *Circular 4/98: Teaching: High Status, High Standards*. London: DfEE.

Ernest, P.: 1989, 'The knowledge, beliefs and attitudes of the mathematics teacher: a model.' *Journal of Education for Teaching, 15*(1), 13-33.

Ernest, P.: 1991, *The Philosophy of Mathematics Education*. London: Falmer Press.

Graber, K. C.: 1996, 'Influencing student beliefs: the design of a high impact teacher education program.' *Teaching and Teacher Education, 12*(5), 451-466.

Gregg, J.: 1995, 'Discipline, control, and the school mathematics tradition.' *Teaching and Teacher Education, 11*(6), 579-593.

Haggarty, L.: 1995, *New Ideas for Teacher Education: A mathematics framework*. London: Cassell.

Joram, E. and Gabriele, A. J.: 1998, 'Preservice teachers' prior beliefs: Transforming obstacles into opportunities.' *Teaching and Teacher Education, 14*(2), 175-191.

Kagan, D. M.: 1992, 'Professional growth among preservice and beginning teachers.' *Review of Educational Research, 62,* 129-169.

Lacey, C.: 1977, *The Socialisation of Teachers.* London: Methuen.

Lave, J. and Wenger, E,: 1991, *Situated Learning: Legitimate peripheral participation.* Cambridge: Cambridge University Press.

Lerman, S.: 1983, 'Problem Solving or knowledge centred: the influence of philosophy on mathematics teaching.' *International Journal of Mathematical Education in Science and Technology, 14*(1), 59-66.

Lortie, D. C.: 1975, *School Teacher: A sociological study.* Chicago: University of Chicago Press.

Nettle, E. B.: 1998, 'Stability and change in the beliefs of student teachers during practice teaching.' *Teaching and Teacher Education, 14*(2), 193-204.

Ofsted: 1993, *Mathematics, Key stages 1,2,3 and 4* London: HMSO.

Ofsted: 1995, *Mathematics, a Review of Inspection Findings 1993/94.* London: HMSO.

Powell, R. R.: 1992, 'The influence of prior experiences on pedagogical constructs of traditional and non-traditional preservice teachers.' *Teaching and Teacher Education, 8*(3), 225-238.

Shulman, L.S.: 1986, 'Those who understand: Knowledge growth in teaching.' *Educational Researcher, 15*(2), 4-14

Simon, M. A. and Schifter, D.: 1993, 'Towards a constructivist perspective: the impact of a mathematics teacher in-service program on students.' *Educational Studies in Mathematics, 25,* 331-340.

Skemp, R. R.: 1971, *The Psychology of Learning Mathematics.* London: Penguin Books.

Smith, D. N.: 1996, 'College ideals and school practice.' *Mathematics Education Review, 7,* 25-30.

Smith, D. N.: 2001, 'The influence of mathematics teachers on student teachers of secondary mathematics.' *Mathematics Education Review, 13,* 22-40.

Smith, J.: 1996, *Getting Started.* Leicester: Mathematical Association.

Thompson, A. G.: 1992, 'Teachers beliefs and conceptions: a synthesis of the research.' In D. A. Grouws (ed.), *Handbook of Research on Mathematics Teaching and Learning* (pp. 127-146). New York: Macmillan.

Zeichner, K. M. and Liston, D. P.: 1996, *Reflective Teaching: An introduction.* New Jersey: Lawrence Erlbaum

THEME 3

MATHEMATICS, LANGUAGE AND MEANING

Recent years have seen a substantial increase in attention to language in the mathematics education research community, both internationally and in the UK. The first three chapters in this section reflect this development, not only by the fact that they all focus in some way on language but also by the enormous diversity in their approaches and subject matter. Each takes as its starting point the idea that a text, whether written or spoken, does something more than simply transmit information.

Chris Bills suggests that examining the ways in which children talk about the mental calculations they perform can provide insight into the ways in which they conceptualise arithmetic. His analysis categorises the metaphors children use, linking these both with their possible mental representations of number and with the representations teachers have provided. He also uses linguistic means to examine the type of generality children construct as they describe arithmetic procedures. His chapter concludes with a hypothesis that different, yet consistent, ways of speaking about arithmetic may indicate distinctive thinking styles.

Many of the mathematics texts encountered by students in the later years of schooling and at university present an impersonal image of mathematics and lack a narrative of human involvement in doing mathematics. By analysing texts and asking what characteristics are essential in written mathematics, the chapter by Candia Morgan suggests that presenting mathematics as an eternal and autonomous system misrepresents the nature of mathematics as the product of human activity. Recognising that mathematical writing does not have to be abstract, formal and impersonal may, it is argued, help to make mathematical writing and mathematical activity more accessible to some of those students who have previously felt excluded by the apparent lack of human presence in mathematics.

How can we explain international differences in attainment in number? In their chapter, Tony Harries, Rosamund Sutherland and Jan Winter suggest that one reason for differences lies in the ways in which children's introduction to number operations is presented in the text books they most commonly use. Focusing on the ways that multiplication is introduced in England and in several other countries, the authors show that the English text books place less emphasis on structure and making links between different representations of multiplication.

The final chapter in this section is also concerned with meanings and with the use of language, but Peter Huckstep and Tim Rowland examine the language used to talk about mathematics rather than that which is used within practices of doing mathematics itself. Questioning claims made in justification of the privileged place mathematics holds in the curriculum, they explore the idea that creativity is, or can be, a feature of school mathematics and conclude that this is a notion that needs to be treated with caution.

8 METAPHORS AND OTHER LINGUISTIC POINTERS TO CHILDREN'S MENTAL REPRESENTATIONS.

Chris Bills

Mathematics Education Research Centre, University of Warwick

The mental representations that 6- and 7-year-old pupils form as a result of their interactions with their teacher's verbal, written, pictorial and concrete material representations has to be inferred from the language they use. In this study many pupils seem to have mental representations which capture surface characteristics of a particular teachers' representation and use metaphoric language associated with that representation when describing their calculations. Pupils' use of 'you' is characteristic of those who adopt a representation-specific procedure, whilst 'if' and 'like' are linguistic pointers to their use of generic examples to describe a procedure. Individual pupils show a preference for the same style of mental representation when describing images and procedures in mathematical and non-mathematical contexts.

INTRODUCTION

I take it as given that pupils' procedural and conceptual knowledge of mathematics is constructed, in part, from experiences with teachers' external representations (written or spoken words or symbols, pictures or concrete objects). I refer to whatever comes into the children's minds, when asked to perform a procedure or recall a concept, as their mental representation. My studies in a primary school have given evidence of language use which indicates that, for some pupils, these mental representations appear to be a re-presentation of their experiences with teachers' representations. When given a mental calculation, then asked how they had performed it, pupils used language which suggests that a particular teachers' representation provides a metaphor for them to communicate how they have calculated. In addition, many pupils described what they had done by prefacing a list of steps, specific to a particular teachers' representation, with the pronoun 'you' ("you add the tens and the ones") indicating that this is a procedure that is generally used. The use of "the tens" rather than the specific digits is taken as further evidence of pupils' use of a generalised procedure.

There is an indication from these studies that some pupils have an ability to express general procedures that is also apparent in contexts other than mental arithmetic. When asked "Tell me how to add twenty" and "Tell me how to write the date" some described what to do in general terms, whilst others illustrated their procedure with a generic example using words such as 'if', 'say' and 'like'. There was also similarity in the type of 'image' description pupil's gave in response to both: "What is the first thing that comes into your head when I say 'hundreds tens and units'?" and a parallel question: "... when I say 'sentence'?"

My hypothesis is that the qualitative differences in pupils' language use is indicative of the qualitative differences in their conceptual constructions and that an individual's progress toward recognition of generality of procedures may be indicated by changing language use. This paper gives a brief literature review and presents evidence on which the hypothesis is based. The analytic tools developed here are intended to be used in a longitudinal study of the development of pupils' mental representations under the influence of their teachers' use of a variety of external representations. Metaphoric language and other linguistic pointers give clues to the influences on pupils' mental representations.

METAPHORIC LANGUAGE AND OTHER LINGUISTIC POINTERS

Whenever we use the language usually associated with one concept to describe a different concept we can be said to be using a metaphor. When we talk, for instance, of 'remainders', 'carry figures', 'take away', etc. we are using language of handling objects in the context of numerical procedures. 'Odd', 'even', 'square', 'large', 'small' are all descriptions of numbers related to physical contexts, 'positive' and 'negative' are words associated with non-mathematical concepts. If we 'find' or 'work out' an answer there are the overtones of searching and making an effort. Some phrases may be metaphoric but connections with other contexts are not recognised – why do we 'round to the nearest' or why is it that six 'goes into' 42? Many classroom phrases do however stem from activities: six 'lots of' seven, 42 'shared by' six, the 'difference between' 19 and 25. We continue to use the words even when not engaged in the physical activity because they have become associated with the symbolic process.

In *Metaphors We Live By*, Lakoff and Johnson (1980) have suggested that the use of metaphor both indicates and shapes our conceptualisations of the real world. Our use of the language of one concept to communicate another indicates that one concept has shaped our understanding of the other. In the context of this paper, the use, by pupils, of the language associated with a teacher's representation is taken to indicate 'the metaphor they calculate by'. It is important to note that this may be no more than a linguistic metaphor in that it provides a language for communicating a way of calculating even though they have no visual image of the teachers' representation. The choice of language may, however, indicate that the teacher's representation provides a cognitive metaphor for doing the calculation, perhaps that a mental analogue of that representation has been used. In Black's view "every metaphor is the tip of a submerged model" (1979, p 31) and Lakoff and Nunez insist that

> Metaphor does not reside in words; it is a matter of thought. Metaphorical linguistic expressions are surface manifestations of metaphorical thought. (1997, p 32)

Johnson (1987) suggests that the origins of conceptual metaphors lie in our bodily experiences; that metaphor is a cognitive structure through which we make use of pattern in our physical experience to organise our more abstract understanding. Our bodily movements and interactions in physical situations are structured by image schemata and this structure is projected by metaphor onto abstract domains. For example, 'balance' is an image schema of patterns common to all situations requiring bodily equilibrium. We experience bodily responses to imbalance. The 'balance schema', he suggests, gives rise to meaning for equivalence relations (symmetry, transitivity and reflexivity) and gives rise to the concept of equality of magnitudes.

Lakoff and Nunez (1997) list three basic 'grounding' metaphors: 'Arithmetic is Object Collection' (numbers are collections of objects and operations are acts of forming collections); 'Arithmetic is Object Construction' (numbers are physical objects and operations are acts of object construction); 'Arithmetic is Motion' (numbers are locations on a path and operations are acts of moving along the path). The first two may be seen as instances of the more general 'Arithmetic is Object Manipulation'. The language of the source concept (e.g., manipulation) is used to communicate ideas about the target concept (e.g., arithmetic) and in so doing may set constraints on understanding of the target concept. Whilst these metaphors, which may be pre-linguistic, are 'grounded' in the sensorimotor experiences of everyday life, they are further developed in the classroom.

Pimm (1995) has also drawn attention to 'manipulation' as the core metaphor for doing mathematics. The manipulation of tangible referents of numbers (adding more counters or taking some away) provide the physical and linguistic metaphors for mathematical operations. Addition becomes synonymous with increasing and subtraction with taking away, thus decreasing. When the metaphor 'Subtraction is Take Away' is the cognitive model for a child, subtraction of negative numbers becomes problematic. Similarly 'Multiplication is Lots Of' and 'Division is Sharing' leave children ill equipped for calculations with anything other than natural numbers. In the same way, manipulation of symbols can provide restrictive metaphors, for example, 'Multiplication By Ten is Adding a Nought'.

The external representations used by the teacher (words, drawings, physical materials, real life contexts) are used with the intention to communicate a mathematical idea. Thus representations are intended to provide metaphors. Concrete-material representations such as Dienes blocks, hundred squares and number tracks, used for place value, which are intended to be 'structure-oriented', are sometimes referred to as 'physical metaphors' (Resnick and Ford, 1981).The representations that teachers use are not the mathematics but a transformation of the mathematics into a communicable form (Kang and Kilpatrick, 1992). It is not surprising, then, that children should use language associated with these representations in order to communicate how they have calculated. Sfard (1994) warns, however, that metaphor must be constructed by the individual and is not simply communicated. She argues that familiarity with a process is a basis for reification and concludes that "reification is the birth of a metaphor". Gray, Pitta, and Tall (1997), point to pupils' individual differences and add another warning that some children's mental representations may merely imitate enactive procedures rather than encapsulate them. There is a danger then that pupils may adopt the language without the understanding of numerical procedures that the materials were intended to illustrate.

One concern of this paper is the metaphoric language which indicates that a mental representation may have been initially formed as a result of experience with material aspects of a teacher's representation. The other interest of the study is the language which children use which might indicate how that mental representation is developing. Rowland (1999) notes that the use of the pronoun 'you', to refer to generalities, is common in non-mathematical situations where 'you' is used in place of the more formal 'one', particularly by children, for instance, in their description of rules of games. He also suggests, however that the use of 'you' is an effective 'pointer' to a quality of thinking. For the pupil in Rowland's study the shift from 'I' to 'you' in a problem solving setting signified her move from working with specific numbers to expressing a generalisation. In an earlier study (Rowland, 1992) he illustrated that the use of 'it' is also a linguistic pointer to a child's attempts at generalisation.

Whilst a generalisation may ultimately be expressed for 'any' number, it is common to choose specific numbers to illustrate a procedure. When particular numbers are chosen simply in order to demonstrate the procedure it is often referred to as a 'generic example' and it may be a step toward a more formal generalisation. The generic example uses a particular number to stand in for a class but doesn't rely on any specific property of that number (Mason and Pimm, 1984). Thus the use of a generic example marks a step toward a statement of generality but is expressed in terms of the specific. Adults often preface a generic example with 'for example' or 'consider for instance' or 'suppose', the children in this study frequently used 'if it's like'

Whilst 'like' would seem to be an obvious linguistic pointer to a recognition of an analogous relation children sometimes use it inappropriately. Piaget (1928) believed that young children have a tendency to "connect everything with everything else" as a result of a comprehensive act of perception which ignores detail. Thus, because of insufficient discrimination, anything can be related to anything and young children suggest inappropriate analogies and inappropriate causal relations. He categorised the majority of young children's spontaneous use of 'because' as a "motive for action" rather than as logical relation between cause and effect. He suggested that children's logic is typified by 'transduction' – proceeding from particular to particular rather than induction (from particular to general) or deduction (from general to particular). He argued that young children can only reason about particular cases but admitted that his studies only showed aptitude for use of language.

Tough (1977) found that some 5 year-old children could give justifications and suggest alternative consequences but children from disadvantaged home backgrounds tended to merely describe objects. Tough suggested that these children lacked experience of using language for reasoning at home and found that they could use the appropriate words when encouraged to do so. Vygotsky (1962) found that 80% of children, in his sample, at both 7 and 9 years were able to correctly complete sentence fragments related to scientific concepts ending in 'because' whilst only 60% of the 7 year-olds could do so with sentences related to everyday concepts. He attributed this to the fact that scientific concepts had been learned in collaboration with teachers. Mercer, Wegerif and Dawes (1999) similarly found that when children were taught to use particular language in reasoning the key linguistic features of the children's talk in small groups, which led to correct answers, were use of 'because', 'I think', 'agree' and long turns at talk.

METHOD

The studies were conducted in a Primary school in a large middle-income village near Birmingham which has pupils in three sets for Mathematics, grouped by achievement, in each year group. In the period October 1997 to July 1998 the preliminary study involved observation of one lesson per week with Set 1 in Year 2 and individual interviews audio taped each term with a sample of pupils. From September 1998 these same pupils, then in Year 3 Set 1 with a new teacher, and also Year 3 Set 2 were observed and interviewed as the first year of a two-year longitudinal study. Results presented in this paper are taken from the Year 2 interviews and interviews conducted with Year 3 Set 2 in September 1998 for the purpose of selecting the sample for a longitudinal study.

In the lessons, notes were taken of the materials used by the teacher, his words and gestures when using them, his 'board work', and the activities he gave to the children. In the interviews children were asked to perform mental calculations similar to the questions they had recently been involved with in lessons, then to talk about what had been in their heads. With the exception of the first interview, all questions were presented verbally and pupils did not use any concrete materials or paper and pencil. The question for the first interview in October was 24 + 53 and this was printed in large font on A5 paper. In March the questions were: What comes next after 379? What is 81 add on 10? What is 247 add on 10? What is 30 take away 4? What is 86p take away 12p? What is 41p take away 13p? After each they were asked how they had worked out the answer. In July they were asked: What is 27 add 5? What is 38 add 10? What is 43 take 7? What is 63 take 10? What is 97 add 30? What is seven threes? What is 3 times 6? What is 4 times 10? What is 12 times 10? What is 17 times 10? What is 27 add 45? What is 43 take 27?

In these first interviews pupils were also asked about their visual images and this aspect is reported elsewhere (Bills, 2000b). In order to explore further the pupils' ability to describe both procedures and images not linked with a specific calculation, the Year 3 Set 2 sample selection interviews also provided an opportunity to pilot four questions in addition to the mental calculations. They were asked "Tell me how to ...", one arithmetic procedure and one non arithmetic procedure, and "What is the first thing that comes in your head when I say ...?" with one mathematical and one non mathematical object. This became the pattern for interviews in the longitudinal study.

RESULTS

Analogy and Metaphor

The class teacher used several representations to demonstrate 2-digit addition and subtraction to Year 2 Set 1 during the course of the year. These included a number track from 1 to 105, a hundred square, Dienes base ten blocks, numeral cards printed with single digits, coins, chanting number sequences and a written algorithm. The pupils practised a representation-specific procedure with each of the materials. They added tens, for instance, by: taking ten steps on a number track; going down a column on a number square; having an extra ten-block; replacing the tens digit with one higher; having an extra 10p coin; saying the next word in the word sequence; adding one to the tens column. The language associated with these activities can be seen to provide a wide variety of metaphors.

In the first interviews conducted in October (after six weeks of teaching) pupils were asked to calculate mentally 24 + 53 (presented on paper). The response of one pupil indicated that she had been influenced by the use of Dienes blocks:

Elspeth   Well you add the tens together then you add the units **because its like** in one hand you have the tens and in the other you have the units

This is the only occurrence, in any of the interviews, of the use of the word 'like' to indicate an explicit analogy drawn with a teacher's representation (In the lessons the teacher took the tens in one hand and the units in the other when using Dienes blocks to illustrate adding two 2-digit numbers).

Others used 'like' for appropriate comparisons:

I   How do you know nine is odd?

Christine Because nine is **like** three and three has a bit sticking up.

I   How do you take off ten quickly though?

Teddy   . um . just **like** adding on ten really

I   How did you work that out? (twelve times ten)

A   You start with another ten, it's a hundred and *ten*, and you add, well its just **like** the ten times table again.

By July, however, many pupils' language gave indications of the influence of a teachers' representations without making specific reference to it. For instance when asked to calculate 27 add 45 (presented orally with no paper or materials available to the pupil):

Terry   I **took** the tens then I added them together and then I **took** the units and added them together

The use of the word 'took' indicates that the handling of the digits separately, even if not necessarily handling the Deines blocks, has become a metaphor.

Pupils also showed a consistency over time in their metaphoric language :

Linguistic pointers to mental representations

	March 81 add on 10	July 38 add 10
Christine	well I, . sort of **ignore** the units for a minute and just added like a ten on.	I would **leave** the un, units and just add on ten
Hannah	because you just add a ten on to so you **keep** the um, the last number then you add a ten onto the un, onto the tens.	Well you **leave** the eight as it is and add one to the three.
Jack	ninety one. If you just go eight , nine then you just make it into a ten and you **put the one on**, ninety one.	.. forty-eight. I know my ten times table and then I just **put the eight on.**
Ann	well, **add a ten would take you to a next column** and the unit would just be the same. You know what it is just saying, instead of saying eight you say nine. **Leave** the one how it was.	Well you just **add another ten onto the tens column**. So three so you count one, two, three and the next one is four it's the columns.

The words 'leave', 'ignore', 'keep' and 'put' were used by the teacher and pupils when working with numeral cards. They handled individual cards, moving them around on their desks or in their hands, as they considered each digit of a multi-digit number separately. They used this same language when working with multi-digit numbers written in their books or on the board and when thinking of the numbers 'in their heads'. Ann, above, made it quite explicit that she was describing what to do as if the numbers were written in place value columns.

The language of manipulation of digits was also apparent when the children were asked "What comes next after three hundred and seventy nine?" and the 'grounding' metaphors can be identified:

Arithmetic is Object Collection

Peter   Well . 379 **and one more** it would be 3 hun and seventy-ten.

Cora   . um I **just add one** to, unit.

Arithmetic is Object Construction

Elaine   So I just **put away** the three um 300 and um and I, 79 went after that comes 80 and then I put the other finger **next I do** three hundred and eighty.

John   ......389 .. I **changed** the ten to the next.

Arithmetic is Motion

Ann   ..... cause **we got to** the nine and I was thinking, I thought it would **go on** to ten.

147

Hannah   because **after** 79 **it goes** 80, so you put, you keep, you um, you save the three, keep the three there and then you know **what's after** 7, 79, and then you put 80.

Procedures.

Many of these pupils, in the responses above, mixed their metaphors. As did Elspeth in the following response. She used object collection ("add one") and object construction ("it has a nought at the end") but she also gave a general rule:

Elspeth   Well when you add, you said that it was seventy nine, and add one on and it equals to seve, ss, eighty because **when you add one on to 9 it has a nought at the end.**

Elspeth's use of 'you' and 'because' are the pointers here to her description of a general rule. Similarly Brian gives a rule which is independent of the specific numbers in the original question. Again 'you' and 'because' feature but also 'like' as pointers to an expression of generality and a recognition of analogy:

Brian   Well you just um well um you don't think about the first number, well, you don't change the first number but you um **you have to change the second two numbers** because it's gone on to a nine and it's **like** going on to a ten.

'Like' was also used by others to illustrate with an example:

I   How do you know if a number is odd or even?

Teddy   Well you, can have, shares **like**, you can share six, **like** if you had two people, one half , give her three and the other half is three, and if you have two people and they have five one person would have to have about three

'Because' was most frequently used when a standard procedure was used as justification for an answer:

I   Right, how you do know those so quickly then?

Dennis   **Because** all you need to do is take, is take away, um, the number that's next after ten.

Others gave more of a sense of causation:

I   How do you know that ninety add thirty is hundred and twenty then?

Terry   ninety add thirty, **because** it's one more ten to get to a hundred and you've got twenty left so I add it on

Hazel's use of 'we', in the following, may be an explicit indication of her acceptance of the representation-specific procedure that has been taught, as well as her acceptance of the language of the representation used in the classroom, but it was rarely used by other pupils.

Hazel   Because um three hundred and seventy nine **we have to change** the um ten, we have to change the 7 to an 8 and **we take 9 to an oh.**

'You' was, however, commonly used by this teacher and these pupils when describing procedures and its use by pupils may be seen as evidence that their mental representation includes the procedure as well as the material. When asked "How did you work that out?" after they had performed a mental calculation, pupils frequently adopted the 'you-mode' even if not giving a general rule as explicitly as Elspeth and Brian.

Hannah explained how she arrived at the correct answer to "What is 45 add 27?" She described what 'you' do, though she made a calculation error when doing so:

Hannah   **You** add the two, **you** add the two to the six, then **you** add the, whatever it was, **you** add the six to the seven makes thirteen then **you** add the one of the thirteen to the six.

Here the representation-specific procedure is that of the written algorithm, expressed in such detail as "the one of the thirteen". Hannah gave just such a description for "What is 43 take 27?", again using 'you', detailed even to the extent of making the common error leading to the answer 24:

Hannah   Well **you** take the two from the four, we got two, . then **you** take the three from the seven, three from seven equals four, so twenty-four

With these more difficult two-digit calculations, however, the use of "you' was less common than when doing the easier questions. This suggests that pupils were expressing confidence in a familiar, non-counting, classroom procedure when using 'you' and otherwise used 'I'. For instance in response to "How did you work that out?" after "What is 45 add 27?":

Malcolm   ....sixty-two? I was counting on um to like um doing forty-five, forty-six, forty-seven, forty-eight,

Terry   .... seventy-two. I started with twenty and added the forty on then I added the seven on then I added the five on.

Those who calculated by a non-representation-specific method always used 'I', for instance "What is 43 take 27?":

Teddy   .... sixteen, I took off the twenty and took off the three and took off another four

'I' and 'you' were occasionally mixed when there was a mixture of known fact and procedure in use:

Cora   **You** just add one, I know nine comes after 8.

When the question is non-routine 'I' may be used for the new thinking and 'you' for routine. For instance going beyond one hundred was not routine for these pupils. So, for "97 add 30":

Cassy   Well I do . I leave the ninety then I add ten on that makes a hundred and then add, the two tens that makes twenty **and then you** add **the** seven

149

The sense that after the hard new thinking Cassy dropped into a routine is further heightened by her use of "the" seven.

Whist 'you' was common when pupils were asked "How did you..." it was almost universally used when asked "How do you ...?" or "How would you ...?". In order to explore pupils' ability to describe procedures the first interviews conducted with Year 3 Set 2 in October 1998 included two questions "Tell me how to add twenty on to any number" and "Tell me how to write the date".

Only two of the twenty pupils interviewed gave a procedure in general terms, for both questions, without reference to any specific number or date:

	... add twenty on to any number	... write the date
Shaun	You just add a ten and then again	. Just like get what the day it is and then put the month and then the years
Bobby	.. um you count how many, ones, um units, to make um the sum and then you add the rest on	. what day it is, the month it is and the year it is.

Notice also that 'you' is used in the arithmetic procedure but not in the un-taught date writing routine.

Over half of the pupils gave a generic example where the particular number seemed only to be used as an illustration. Linguistic pointers to this are 'if' and 'like':

	... add twenty on to any number	... write the date
Jonathan	. well if there was like 47 or something like that then you knew it would be sixty something so you would just go 1, 2, 3, 4, 5, 6, 7, 8, 9, 10, on your fingers and do it again	if it was like September the 6th I'd write, and um 98, You'd write 6 dot ,nine nineteen ninety eight
Geoff	. um, . you'll like have 20, then you have 5 there and then you just like 20 and then you add on like 5 if you were ...	um, . well you write, well if it's the 20th of October then you write 20 and then t, h and then um, . c, t, o, . o, b, e,r
Steve	Well if you're going to add up to twenty, uh, 60, you add the tens like twenty, (mutters) 4 more tens, you've only got 2 and you had 4 more tens, you have 60	Well in maths you always write the short date, like today's date 20 dot 10 dot 98

Some pupils could only answer by saying a specific date though none performed a specific addition. Only three pupils could give no response.

Types of Image.

The Year 3 Set 2 pupils were also asked "What is the first thing that comes into your head when I say 'hundreds tens and units'?" and "What is the first thing that comes into your head when I say 'sentence'?" All but three of them were consistent in the type of image projected. The majority gave a specific response for both. For instance:

	... 'hundreds tens and units'	... 'sentence'
Pam	... 2 hundred, and sixty, seven.	I jumped on a cat.

Three, however, talked of general or procedural aspects for both:

	... 'hundreds tens and units'	... 'sentence'
Bobby	..... I know how many um ones in tens, in hundreds.	um, **you** have to have a doing word in a sentence, a verb, um, and **you** have to have, um, a capital letter at the beginning of a sentence and a full stop at the end.
Joe	Well just put, the hundreds, but it has to be over a hundred or (incomprehensible) and then **you** have to do tens, and then **you** just put the um ones next.	well, it's, **you** just write down um, and **you**, write lots of words, until like **you** get to a place where **you** can put a full stop, to make a sentence.
Hester	. 5 hundred and 3, and I know one's really high and one's really low.	**You** write words in sentences.

Some gave responses using the linguistic pointers associated with generic examples:

	... 'hundreds tens and units'	... 'sentence'
Steve	. **you** um, first thing that came in my head was **you**'ve got three boxes, and **if you** have to copy out a number **like** 937, **you** do 9 in the first box, 3 in the second and 7 in the last box	... well **if you**, when **you**'re reading books, **if you**'re. Sentences are **like**, well, hard to explain, . something, **You** don't have read them, if there are no full stops, with it **you**'d run out of breath, always have a full stop at the end, well uh,. the idea is to stop **you** running out of breath.
Robert	Well I get hundred, I get **like** one, oh, oh, hundred, I get , I mean I get ten, and I get, no, I get 5 then ten then a hundred.	it's **like** doing sentences, you've got **like**, it's not **like**, "day whe," it's not it's **like** "day", **like** that, it's meant to be "one day" like "One day a boy walked the street" It's **like** that.

151

DISCUSSION

The evidence presented suggests that the language of a teacher's representation and the procedures associated with it provide pupils with a metaphor for communicating their methods of calculation. It is hypothesised that the language that pupils use is an indicator of the mental process they have employed for a calculation and consequently that their qualitatively different language is indicative of qualitatively different conceptualisations. The use of metaphoric language is an indication of the influence of the external representation on the internal representation. For most of these pupils this language is the only manifestation of their mental representation because they do not claim to have any visualised mental image.

The examples of metaphoric language given in this paper could be dismissed as mere 'figures of speech' but there is a consistency amongst pupils that suggests they have been influenced by the common experiences. There is some confirmation also when the language is an accompaniment to a mentally visualised image. In a case study of Elspeth I have illustrated that the language she uses is consistent with an image of the teachers' representations that she claims to 'see' in her head (Bills, 2000a).

The use of 'I' and 'you' may only distinguish between construction (own mental representation) and re-construction ('copied' teachers' representation, merely mimicking the teachers' style of speech). There are strong indications, however, that pupils use 'I' in circumstances where they are less confident or using personal methods and 'you' when they are using familiar and taught routines. Thus for some pupils the change from one to the other may also mark a transition from working with specific examples to recognising generality. 'You' was used by those pupils who gave descriptions of general rules. The use of 'you' could thus be a pointer that a general procedures is in use. Its use could indicate that these pupils' re-present a general procedure to themselves when asked to perform a calculation.

Some pupils have given evidence of an ability to express procedures in general terms, or to give generic examples to illustrate procedures, both in an arithmetic and in a non-arithmetic context. There is evidence also of a preference for different types of description when asked "What is the first thing that comes into your head ...?". These may also be termed specific, generic and general and they have the same linguistic pointers associated with them. This consistency in style of description of procedures and descriptions of images may be indicative of distinctive thinking styles.

CONCLUSION

I have given examples elsewhere (Bills, 2000b) of the relatively infrequent use of mental visual images by Year 2 pupils when making mental calculation. I have suggested, however, that their mental representations are 'image like' in reproducing aspects of their experiences with their teachers' representations (Bills and Gray, 1999). The language that is used by individual pupils is sufficiently different for me to contend that these mental representations are the children's own constructions yet sufficiently similar to suggest that they have a common source in the teacher's representations.

The linguistic pointers suggested in this paper will give indications of the developing conceptual styles of pupils in an analysis of the influence of teachers' representations on the two samples of pupils over a two year period. An analysis will be made of the children's changing, or invariant, use of metaphor and the degree to which they express generality when describing both procedures and images.

REFERENCES

Bills, C. J.: 2000a, 'The role of re-presented experience in mental calculation: A case study.' In A. Gagatsis and G. Makrides (eds.), *Second Mediterranean Conference on Mathematics Education*, Vol. 2 (pp. 290-299). Nicosia, Cyprus.

Bills, C. J.: 2000b, 'The influence of teachers' presentations on pupils' mental representations.' In T. Rowland and C. Morgan (eds.), *Research in Mathematics Education Volume 2: Papers of the British Society for Research into Learning Mathematics* (pp. 45-60). London: British Society for Research into Learning Mathematics.

Bills, C. J. & Gray, E. M.: 1999, 'Pupils' images of teachers' representations.' In O. Zaslavsky (ed.), *Proceedings of the 23rd Conference of the International Group for the Psychology of Mathematics Education*, Vol. 2 (pp. 113-120). Haifa.

Black, M.: 1979, 'More about Metaphor.' In A. Ortony (ed.), *Metaphor and Thought* (pp. 19-43). Cambridge: Cambridge University Press.

Gray, E., Pitta, D. & Tall, D.: 1997, 'The nature of the object as an integral component of numerical processes.' E. Pehkonen (ed.), *Proceedings of the 21st Conference of the International Group for the Psychology of Mathematics Education*, Vol. 1 (pp. 115-130). Lahti.

Johnson, M.: 1987, *The Body in the Mind*. Chicago: University of Chicago Press.

Kang, W., and Kilpatrick, J.: 1992, 'Didactic transposition in mathematics textbooks.' *For the Learning of Mathematics, 12*(1), 2-7.

Lakoff, G. and Johnson, M.: 1980, *Metaphors We Live By*. Chicago: The University of Chicago Press.

Lakoff, G. and Nunez, R. E.: 1997, 'The metaphorical structure of mathematics.' In L. D. English (ed.), *Mathematical Reasoning: Analogies, Metaphors, and Images* (pp. 21-89). Mahwah,NJ: Lawrence Erlbaum Associates.

Mason, J. and Pimm, D.: 1984, Generic examples: Seeing the general in the particular. *Educational Studies in Mathematics, 15*(3), 277-289.

Mercer, N., Wegerif, R., and Dawes, L.: 1999, 'Children's talk and the development of reasoning in the classroom.' *British Educational Research Journal, 25*(1), 95-111.

Piaget, J.: 1928, *Judgement and Reasoning in the Child*. London: Routledge and Kegan Paul.

Pimm, D.: 1995, *Symbols and Meanings in School Mathematics*. London: Routledge.

Resnick, L. B. and Ford, W. W.: 1981, *The Psychology of Mathematics for Instruction*. Hillsdale: Lawrence Erlbaum Associates.

Rowland, T.: 1992, 'Pointing with pronouns.' *For the Learning of Mathematics, 12*(2), 44-48.

Rowland, T.: 1999, 'Pronouns in mathematics talk: Power, vagueness and generalisation.' *For the Learning of Mathematics, 19(2)*, 19-26

Sfard, A.: 1994, 'Reification as the birth of metaphor.' *For the Learning of Mathematics, 14*(1), 44-55.

Tough, J.: 1977, *The Development of Meaning: A Study of Children's Use of Language*. London: George Allen & Unwin.

Vygotsky, L.S.: 1962, *Thought and Language*. Cambridge, Mass: The M.I.T. Press

9   A TRANSNATIONAL COMPARISON OF PRIMARY MATHEMATICS TEXTBOOKS: THE CASE OF MULTIPLICATION

Rosamund Sutherland and Jan Winter,
Graduate School of Education, University of Bristol

Tony Harries, University of Durham,

This paper centres around a transnational comparison of primary mathematics textbooks. The focus is on the similarities and differences between the ways in which images related to multiplication are represented in textbooks. The results of the study show that the English and USA texts analysed place less emphasis on mathematical structure and the linking of mathematical representations than the texts from France, Hungary and Singapore.

INTRODUCTION AND BACKGROUND

We suggest that mathematics textbooks emerge from the mathematics education culture of a particular country, in that they are likely to reflect the ways in which textbooks writers view both mathematics and the teaching and learning of mathematics. In this respect, a comparison of textbooks provides an insight into similarities and differences between the ways of viewing mathematics education across different countries.

This chapter derives from a transnational comparison of primary mathematics text books [1]. In analysing text books we were influenced by a view that pupils' construction of knowledge cannot be separated from the multifaceted external representations of this knowledge which envelope the learning pupil. These include spoken language, gesture and written representations, and include both informal representations (for example counting on fingers) and formal representations (for example multiplication algorithms). Representations also include pictures, icons and such mathematical symbols as tables, graphs and arithmetic symbols. Some representations are part of a more general mathematical culture (for example long multiplication), some are part of 'everyday' culture (for example tallies on paper) and others have been developed for pedagogic purposes (for example the empty number line). We call these latter representations pedagogic representations.

We suggest that external representations are potentially transformative, in that they enable a person to do something that he/she could not do alone. For example, tallies on paper support an individual to count a large number of objects while the long multiplication algorithm enables an individual to multiply numbers that would be very difficult without such a paper-based algorithm. This perspective derives from the work of Vygotsky (1978) and Wertsch (1991) and emphasises that human action is mediated by 'technical' and 'cognitive tools'. External representations have the potential for both enhancing and constraining learning. The long multiplication algorithm, for example, could enhance an individual's calculating power, but it could also constrain if introduced inappropriately early in the process of learning to multiply. As Kress and van Leeuwen point out, unschooled children have both more and less freedom of expression:

> more because they have not yet learned to confine the making of signs to the culturally and socially facilitated media, and because they are unaware of established conventions and relatively unconstrained in the making of signs; less, because they do not have such rich cultural semiotic resources available as do adults (Kress & van Leeuwen, 1996, p. 7).

There has been very little research focusing on the ways mathematics is presented in textbooks and the likely effects on the construction of meaning. This may be because many mathematics educators support what Voigt (1998) calls the folk belief that tasks, questions and symbols of mathematics lessons have definite, clear-cut meanings. Or it may be because of a view, often expressed by mathematics educators, that 'good' teachers do not use mathematics textbooks.

One area in which there has been some research is that of the use and interpretation by children of illustrative material. Recent work of Santos-Bernard (1997) suggests that children do not necessarily read and use illustrations in the same way as they are read by authors and teachers. Santos-Bernard analysed the approach of pupils in Mexico and the UK to illustrations in mathematics texts and found that pupils may over-interpret the information presented in an illustration, or alternatively may ignore pictorial information when it is pertinent to the problem being solved. Low attaining pupils, in particular, found it hard to extract appropriate information from both illustrations and text and were far more likely to misinterpret pictorial information than other pupils. Voigt has also drawn attention to the fact that ambiguity and negotiation of meaning are essential features of mathematics classrooms: "Every true-to-life empirical situation (given as a story, picture, text, etc) can be mathematised in various ways, depending on one's interest" (1998, p. 208). He presents evidence that children interpret real-life pictures in text books in many different ways although he says that "many authors of textbooks as well as mathematics teachers seem to assume that such pictures have unambiguous mathematical meanings and represent tasks that have definite solutions"(*ibid.*, p.196).

Our aim in this chapter is to focus in detail on the ways in which mathematical ideas are represented in text books, and for this purpose we restrict our attention to the introduction of multiplication within a sample of textbooks (and associated teachers' guides) from England, France, Hungary, Singapore and the USA [2]. We do not claim that the textbooks analysed represent all the textbooks that might be used in a particular country. However it is interesting to note that, as with the TIMSS analysis of classroom processes (Steigler and Hiebert, 1997), when we analysed two text book schemes for a particular country (in the case of England and France) we found that there were considerable similarities between these texts. In these cases, we have chosen to use only one text in what follows.

Our analysis of texts has been framed by a consideration of:

- The nature of the images with which pupils engage as they read the text, this includes pictures, diagrams and symbols. Analysis of these images has been influenced by the following categories which derive from the work of Botsmanova (1972):

  — object-illustrative images, which illustrate objects in the problem but not the mathematical structure of the problem;

  — object-analytical images, which reflect the mathematical structure of a problem;

  — abstract spatial diagrams, which reflect abstract numerical relationships.

- The ways in which pupils are introduced to links between mathematical concepts and the role of images in this respect.

REPRESENTING MULTIPLICATION

The affordance of different representations

In all of the textbooks analysed the notion of multiplication is first introduced as repeated addition of equal groups, and in the English, French and Hungarian texts this evolves into representing multiplication as an $m \times n$ array. The Hungarian, French and English texts also use the notion of function in the form of 'jumps on a number line' to introduce multiplication. Whereas we are aware that there are arguments for the advantages and disadvantages of each approach (see, for example, Davydov, 1991; Anghileri, 1991) our focus is on analysing how the ideas are represented on the text book page. Interestingly the teachers' guide to the French text book scheme is the only one which explicitly addresses the advantages and disadvantages of each approach (Brégeon, Flouzat, Dossat, and Vicen, 1996). In this guide, teachers are told that presenting multiplication as repeated addition has the advantage that it can develop as a natural extension of addition. They are also told that multiplication as repeated addition has disadvantages, in that it does not lead directly to the notion of commutativity, and is not extendible to the product of two decimal numbers. They are told that presenting multiplication as 'an array' (a) emphasises the commutativity of a product, (b) allows multiplication calculations to draw on the distributativity of multiplication over addition, e.g., $12 \times 23 = (10 + 2) \times (20 + 3) = 200 + 30 + 40 + 6$, and (c) can be extended to decimal numbers. Later in this teachers' guide when the number line is being introduced, teachers are told that there are many situations leading to calculating a product that cannot be represented by an 'array diagram' but can be represented by a 'function diagram' (an example of such a situation is problems where you have to find the price of n articles knowing the price of one article). This French teachers' guide explains that the number line is particularly useful to support mental calculations and also offers a first entry point to memorising multiplication tables. It could be argued that this type of analysis of the 'affordances'

(Gibson, 1979) of a particular representation from the point of view of learning mathematics is an important aspect of teachers' pedagogic knowledge.

Singapore and Hungary – use of multiple representations which reflect structure

The teachers' guide to the Singapore texts explicitly states that the aim is to present pupils with concrete, pictorial and abstract approaches to mathematics and this can be seen in the first introduction to multiplication (Figure 1).

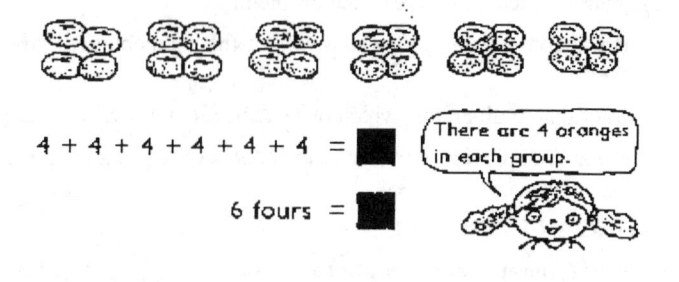

Figure 1: Introduction to multiplication (Singapore text, age 6-7)

Here pupils are presented with an image of equal groupings of 'real' objects, talking heads which represent the image in words "There are 4 oranges in each group", a number sentence using symbols "$4 + 4 + 4 + 4 + 4 + 4$", and a hybrid sentence using numbers and words "6 fours". These representations emphasise mathematical structure and the links between the different ways of representing multiplication. The multiplication symbol is introduced a few pages later as 'This is multiplication. It means putting together equal groups'. Pupils are supported to see the equivalence of, for example, $4 + 4 + 4 + 4 + 4 + 4$ and 6×4, with these representations placed next to each other on the page. The emphasis in the Singaporean texts is on the whole/part relationship (also used later when introducing division). Thus in the Singaporean texts, multiplication is about putting together equal groups and division is the inverse of this. The teachers' guide also uses an image to represent the part-whole principle (Figure 2). The Singaporean teachers' guides make extensive use of object-analytical and abstract spatial diagrams when discussing how to introduce a mathematical idea to pupils (the French teachers' guide is similar in this respect). The illustrations in the Singaporean texts are almost entirely of the object-analytical type. The pictures represent the idea of equal groups and the words explain the pictures in terms of the mathematical structure being illustrated. Diagrams and pictures are focused on illustrating the mathematical concepts being presented, and the mathematical structure of these ideas. Decorative illustration is minimal – there are small pictures of dinosaurs at the start of each chapter and occasionally there are pictures that illustrate the context within which the concept is being explored.

Unit 4: Multiplication

① Adding Equal Groups

Textbook 1B Pages 42 to 45
Workbook Exercises 30 to 32

Instructional Objectives

• To recognise equal groups and find the total number in the groups by repeated addition.
• To use the mathematical language such as '4 threes' and '2 groups of 5' to describe equal groups.

Notes for Teachers

When a whole is made up of two parts,

For example,

• we *add* to find the whole given the two parts;
• we *subtract* to find one part given the whole and the other part.

$4 + 3 = 7, 3 + 4 = 7$
$7 - 3 = 4, 7 - 4 = 3$

The part-whole concept of addition and subtraction can be extended to multiplication and division.

When a whole is made up of equal parts,

For example,

• we *multiply* to find the whole given the number of parts and the number in each part;
• we *divide* to find the number in each part given the whole and the number of parts; or
• we *divide* to find the number of parts given the whole and the number in each part.

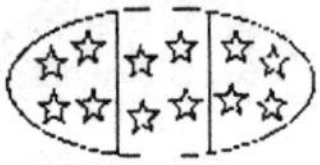

$4 \times 3 = 12, 3 \times 4 = 12$
$12 \div 3 = 4, 12 \div 4 = 3$

Figure 2: The part-whole principle (Singapore teachers' guide, age 6-7)

Szorzás

Peti 3 doboz bonbont vett. Minden dobozban 5 bonbon volt.
Hány bonbont vett összesen?

$5 + 5 + 5 = \boxed{15}$
$3\text{-szor } 5 = \boxed{15}$

Rövidem így irhatjuk:
$\boxed{3} \cdot \boxed{5} = \boxed{15}$

Róka 5 doboz bonbont vett. Mindem dobozban 3 bonbon volt.
Hány bonbont vett összesen?

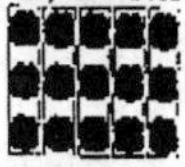

$3 + 3 + 3 + 3 + 3 = \boxed{15}$
$5\text{-ször } 3 = \boxed{15}$

Rövidem így irhatjuk:
$\boxed{5} \cdot \boxed{3} = \boxed{15}$

A szorzásban a tényezők felcserélhetők: $\boxed{3} \cdot \boxed{5} = \boxed{5} \cdot \boxed{3}$
Ezért ezt $\boxed{3 \cdot 5}$. Igy olvashatjuk: 3-szor 5,
3 szorozva 5-tel,
a 3-nak 5-szöröse stb.

Figure 3: Introduction to multiplication (Hungarian text age 7-8)

When multiplication is introduced in Hungary (Figure 3) the emphasis (as in Singapore) is on putting equal groups together to make a whole and splitting a whole up into equal parts (sometimes with remainders). As in the French and Singaporean texts different ways of representing multiplication are all linked together, with pictures, words and symbolic representations appearing together. Images in the Hungarian texts are almost entirely of the object-analytical type, with no use of purely decorative images. Words are used to explain the diagrams/pictures, and there is no use of talking heads.

The English text – a focus on images as illustrations

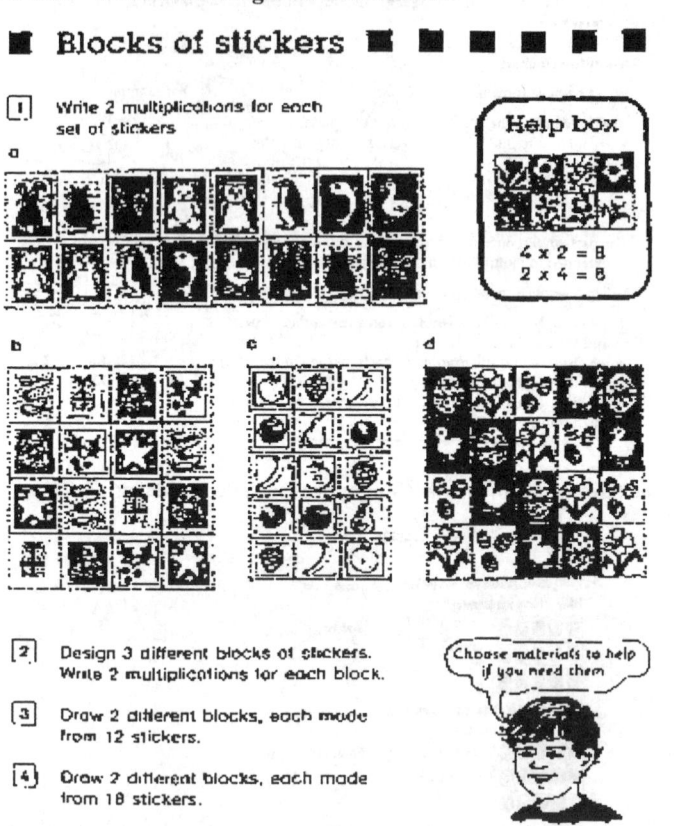

Figure 4: Multiplication: arrays (English text, age 6-7)

If we compare the approach used in the Singapore text to the approach used in the English text we find a number of differences. In the English text, there is less emphasis on mathematical structure with respect to the way in which images are represented on the page and the images are mainly of the object-illustrative type. When multiplication is first introduced in the English text images of equal groups are presented together with a symbolic representation, such as for example $4+4+4 = ...$ The multiplication symbol ($\times$) is introduced much later. This separation of a standard mathematical symbol from a mathematical concept being introduced is a characteristic of the English texts analysed, whereas the French, Singaporean and Hungarian texts all introduce standard mathematical symbols at almost the same time as a concept is being introduced.

The notion of multiplication as an array is introduced in the English texts approximately a year after the first introduction of multiplication, as a way of introducing the idea of commutativity and other properties of multiplication. In the textbook page illustrated in Figure 4 there is very little emphasis on mathematical structure and the linking of representations. The title of the page is "Block of stickers" and this seems to be where the emphasis lies. We suggest that the images used on this page are mainly of the object-illustrative type and that these illustrations are likely to distract pupils from noticing the commutative property of multiplication.

The array notation is also used in the English text to illustrate the idea of decomposing the product of two numbers (Figure 5). In this English textbook page there is an inconsistency between the notation used to represent the 10 rows of 5 stickers ($5 \times 10 = 50$) and the 2 rows of 6 dots in the help box ($2 \times 6 = 12$). This could cause considerable difficulties for pupils who are struggling to make sense of this notation system and are not yet confident with the idea of commutativity.

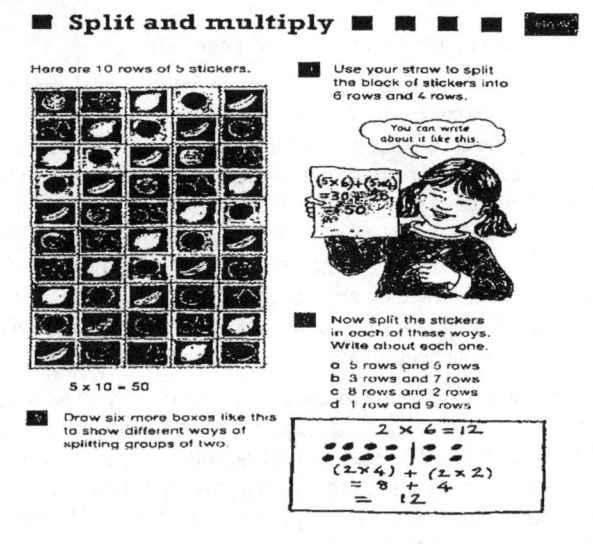

Figure 5: Decomposition of a product (English text age 7-8)

On the English textbook page, there is very little support for pupils to link the numeric representation of decomposition to the array notation. Compare this with the introduction of a similar idea on a French textbook page (Figure 6). Here object-analytical images are used which make links between the numeric representation and the decomposition of the array. Pupils are explicitly told to use the array to help them calculate 15×12. This product is one that would be difficult to calculate without the use of the array, in contrast to the example used on the English text book page (2×6).

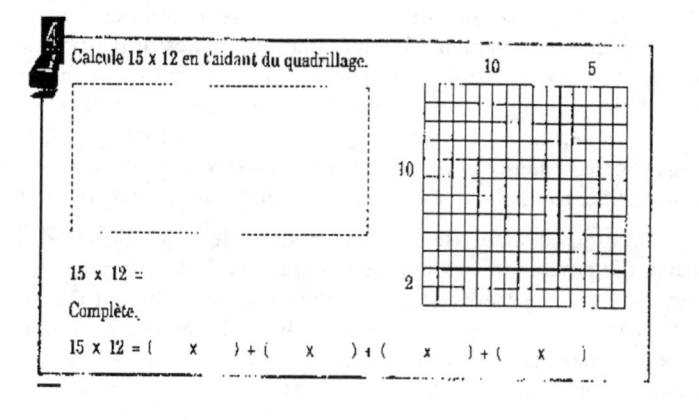

Figure 6: Decomposition of a product (French text age 7-8)

We wonder if the lack of emphasis on mathematical structure and the linking of mathematical representations in the English text relates to the production process, with designers making final decisions about the ways in which images are presented. Even if this is the case it is interesting that English text book writers allow designers to work in this way, because it suggests that they do not believe that the way in which a text book page is presented makes any difference to the mathematical meanings constructed by pupils. This seems to reflect a view that mathematical objects exist almost independently from how they are represented, that is, a quasi-platonic view of mathematics. This quasi-platonic view contrasts with the constructivist perspective which is alluded to in the teachers' guide to the English text book scheme.

This apparent lack of attention to the mathematical detail of how an idea is presented in the English texts can also be seen in the images that are used to introduce the number line. The activity in the English text is called 'Space Jumps' whereas an activity for pupils of a similar age in the French text is called 'The product of two numbers'. Both texts use the idea of 'jumps' on a number line but the English text emphasises a physical activity with an image of a space 'person', presumably about to jump in space. There is no such image in the French text, where the activity is a more abstracted one of working on number relationships. The introduction, in the English text, of a spurious 'real' activity is characteristic of both the English text book schemes analysed, and is not evident in the other texts analysed. Texts from the other countries make extensive use of 'real' situations, but these are situations that are genuinely being mathematised. We suggest that the number line is a pedagogic representation used to introduce ideas that relate to mathematical structure leading to the introduction of multiplication tables. Relating the number line to a made-up real situation is entirely artificial and is likely to confuse low attaining pupils.

Representations which scaffold the introduction of a standard algorithm

A characteristic of both the Hungarian and the French texts is the way in which pedagogic representations are transformed over time in order to introduce a standard algorithm (this is also evident to some extent in the Singaporean text). This is illustrated by the introduction of a multiplication algorithm in the French texts. As we have discussed already, the $m \times n$ array is used to introduce ways of decomposing a product of two numbers (Figure 6). This representation evolves over a period of two years as a way of 'scaffolding' the introduction of an algorithm for vertical multiplication (see Figures 7 & 8). In Figure 6, the $m \times n$ array is used to draw attention to the product of the decomposed parts of two 2-digit numbers:

$$13 \times 21 = (10 + 3) \times (20 + 1) = 10 \times 20 + 10 \times 1 + 3 \times 20 + 3 \times 1.$$

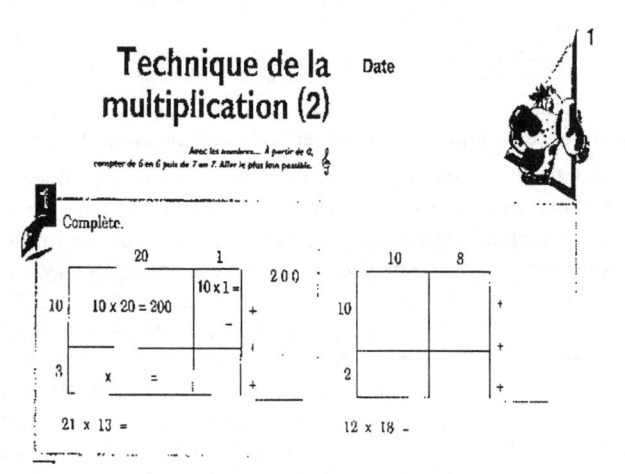

Figure 7: Array representation leading to vertical multiplication algorithm (French text age 8-9)

In Figure 8, there is a shift of attention from the $m \times n$ array used as a 'tool' to support multiplication towards the relationship between this representation and a vertical algorithm representation. This is illustrated with a product that 'needs' support from a vertical algorithm (217×46). We are interested in how French school children are able to make sense of the very detailed image presented in Figure 8 and suggest that research in this area would have to take account of their whole experience of engaging with such images in primary school.

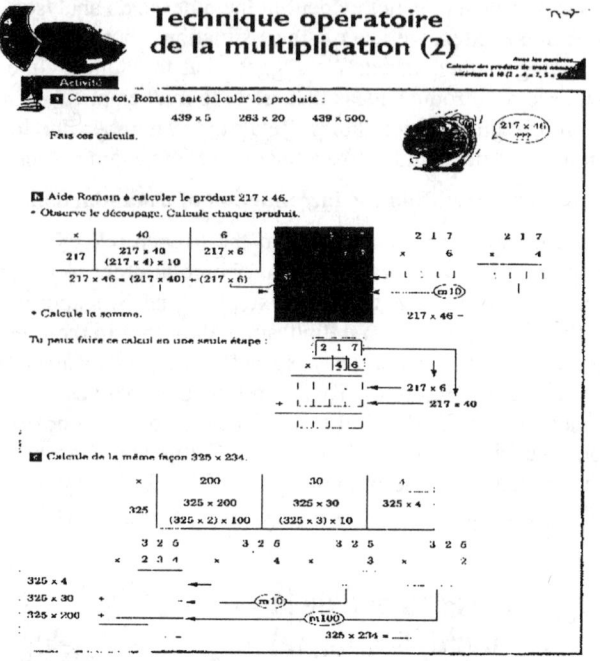

Figure 8: Array representation leading to vertical multiplication (French text age 8-9)

Although in the English text there is also an attempt to introduce pupils to the relationship between the representations of multiplication in a $m \times n$ array and the representation in a vertical algorithm, the images used do not attempt to scaffold the pupils in a systematic way from the use of one representation to another. As we have discussed elsewhere (Sutherland, 1998), the emphasis in the English texts is on diversity of approaches with no particular emphasis on one standard algorithm. The text analysed from the USA is similar to the UK text in that it does not present pupils with images of linked mathematical representations and does not develop representations to 'scaffold' pupils' developing understanding. Of course we do not know how a teacher who is using these texts represents mathematical ideas to pupils (on, for example, a shared white board), but the USA and English texts do not appear to support teachers to develop different ways of representing mathematics in their teaching.

SOME CONCLUDING REMARKS

Our theoretical perspective suggests that the activities and experiences from which pupils construct meaning are related to the materials and resources with which they are presented. In other words, the images pupils engage with will make a difference to what they learn (perception and cognition being intertwined). Our analysis suggests that the Hungarian, French and Singaporean texts books analysed support pupils to develop mathematical meaning by moving along a chain of abstraction from a 'real' situation, to images of this real situation, to mathematical representations of these images of real situations. In contrast, it appears that the English text analysed does not support pupils to construct mathematical meaning in this way. The English text places considerable emphasis on the use of object-illustrative images, probably as a way of relating mathematics to 'realistic' situations. These images are often spurious and sometimes introduce an apparently 'realistic' situation that is presented as if it is being mathematised, but is in fact being used only metaphorically. In other words, the image draws attention to something that has no relevance to the 'mathematics to be learned'. This is different, say, from an image of equal groups of 'real objects' when multiplication is being introduced. It seems that the designers of the English textbook are not aware of these subtle differences. We suggest that low attaining pupils are the ones who are most likely to be distracted by such a 'mathematically incoherent' use of images (as the results of Santos-Bernard, 1997 suggest).

Other characteristics of the English text book scheme which are different from the Hungarian, French and Singaporean schemes are:

- A separation of the introduction of standard mathematical symbols from the first introduction of a mathematical idea.

- Very little emphasis on mathematical structure.

- Very little emphasis on linking mathematical representations.

We recognise that it is often argued that 'good' teachers should not depend on using text books, but we are interested in why there are such substantial differences between the ways in which mathematics is represented on text book pages in different countries. We believe that the results of this study provide a starting point for a more extensive transnational comparison of the practices of teaching and learning mathematics in primary classrooms. We also need to unpack the theoretical influences that influence textbook writers. The related teachers' guides suggest that the Singaporean scheme draws on the work of Bruner (1970) and the French scheme draws on the work of Brousseau (1999) and Vergnaud (1990). Discussion with colleagues in Hungary suggests that Hungarian mathematics education is influenced by the work of Polya (1973).

It is often assumed that mathematical notions such as multiplication are culture free. Our comparative analysis of primary mathematics textbooks puts into question this assumption and suggests that pupils in different countries are likely to construct different meanings of multiplication. In future research we shall be interested in asking questions such as 'does a Hungarian child who has worked with a Hungarian text develop different understandings of multiplication from an English child who has worked with an English text?'. This is not the same as studying differences in attainment as reflected in international studies such as TIMSS. We are interested in understanding the role which cultural tools such as textbooks and the language of the classroom play in the processes of meaning construction. We suggest that this understanding should form the bedrock of the pedagogic knowledge of primary mathematics teachers.

NOTES

1. The project, funded by the QCA, compared textbooks from England, France, Hungary, Singapore and the USA. A parallel study was also carried out by Geoffrey Howson, comparing texts from Canada, Ireland, The Netherlands, Japan, Switzerland and England. Results from these studies are presented in Howson, Harries and Sutherland (1999) and Harries and Sutherland (1999).

2. In order to choose appropriate texts we followed the advice, experience and expertise of teacher educators in each country. The following is an overview of how these choices were made: England – two text book schemes were analysed, chosen from those which are most frequently used by schools; France – two text book schemes were analysed, chosen for us by Bernard Cornu; Hungary – one text book scheme was analysed, chosen for us by an education expert; Singapore – we analysed the one text book used in the country; USA – we analysed a scheme chosen for us by experts at the University of Chicago. For the purposes of this chapter we shall consider only one of the two text books analysed from France and England.

REFERENCES

Anghileri, J.: 1991, 'The language of multiplication and division.' In K. Durkin and B. Shire (eds.) *Language in Mathematical Education: Research and Practice* (pp. 95-104). Milton Keynes: Open University Press.

Brégeon, J. Flouzat, A. Dossat, L. and Vicen, P.: 1996, *Collection Diagonales, Math en Herbe*, Livre du Maître. Paris France: CE1, Nathan.

Botsmanova, M.E.: 1972, 'The forms of pictorial aids in arithmetic problem solving.' In J. Kirkpatrick and I. Wirzsup (eds.) *Soviet Studies in the Psychology of Learning and Teaching Mathematics* (Vol. VI, pp. 105-110). Chicago, USA: University of Chicago.

Brousseau, G.: 1997, *Theory of Didactical Situations in Mathematics.* Dordrecht: Kluwer Academic Publishers.

Bruner, J.S.: 1970, 'Some theorems on instruction in reading.' In E. Stone (ed.) *Educational Psychology.* London: Methuen.

Davydov, V.V.: 1991, 'A psychological analysis of multiplication.' In L. P. Steffe (ed.) *Psychological Abilities of Primary School Children in learning Mathematics*, Soviet Studies in Mathematics Education Vol. 6, (pp. 9-85). Reston, Virginia: National Council of Teachers of Mathematics

Gibson, J.J.: 1979, *The Ecological Approach to Visual Perception*. Boston, USA: Houghton Miflin.

Harries T. and Sutherland R.: 1999, 'Primary school mathematics text books. An international comparison.' In I. Thompson (ed.) *Issues in Teaching Numeracy in Primary Schools* (pp. 51-66). Buckingham: Open University Press.

Howson, G., Harries, T. and Sutherland, R.: 1998, *A Comparison of Primary Mathematics Text Books from Five Countries with a Particular Focus on the Treatment of Number*. Summary Report to Qualifications and Curriculum Authority, London.

Kress, G. and van Leeuwen, T.: 1996, *Reading Images: The Grammar of Visual Design*. London: Routledge.

Polya, G.: 1973, *How To Solve It: A New Aspect of Mathematical Method*. Princeton, New Jersey: Princeton University Press.

Santos-Bernard, D.: 1997, *The Use of Illustrations in School Mathematics Text Books – Presentation of Information*. Unpublished PhD thesis, University of Nottingham.

Steigler, J.W. and Hiebert, J.: 1997, *The TIMSS Videotape Classroom Study*. Washington D.C: National Center for Education Statistics.

Sutherland, R.: 1998, *Mathematics Education: A Case for Survival*. Inaugural Professorial Lecture, May 14th 1998, Graduate School of Education, University of Bristol.

Vergnaud, G.: 1990, 'Problem-solving and concept formation in the learning of mathematics.' In H. Mandl *et al.* (ed.) Learning and Instruction: European Research in an International Context, Vol. 2(2). Oxford: Pergammon Press.

Voigt, J.: 1998, 'The culture of the mathematics classroom: Negotiating the mathematical meaning of mathematical phenomena.' In F. Seeger, J. Voigt and U. Wascescio (eds.) *The Culture of the Mathematics Classroom*. Cambridge: Cambridge University Press.

Vygotsky, L.: 1978, *Mind in Society*. London: Harvard University Press.

Wertsch, J.: 1991, *Voices of the Mind; A Sociocultural Approach to Mediated Action*. London: Harvester.

10

MATHEMATICS AND HUMAN ACTIVITY: REPRESENTATION IN MATHEMATICAL WRITING

Candia Morgan

Institute of Education, University of London

The prevailing image of mathematical writing is formal and impersonal, reflecting an absolutist view of the nature of mathematics. This paper challenges this image. Analysis of texts written by research mathematicians shows that they include different kinds of representations of the human activity of doing mathematics. Genres of school mathematics writing are discussed in the light of this analysis and it is suggested that knowledge of the ways in which mathematicians include their problem solving activity in their writing might be used to support students as they learn to write reports of their own investigative work.

INTRODUCTION

When asked to describe the main characteristics of mathematical writing, most people are likely to respond with terms such as *abstract, formal, impersonal* and *symbolic*. Academic writing in the humanities and social sciences has been affected to some extent by attempts to incorporate the author's human voice (by, for example, using the first person singular and avoiding use of the passive) [1]. Even among mathematics academic research papers there are some that give their readers explicit insights into their authors' personal motivations and actions (Burton and Morgan, 2000). Yet the prevailing image of mathematical writing, perpetuated in most of the texts encountered by students in the later years of schooling and at university, is still impersonal, lacking a narrative of human involvement in doing mathematics. Indeed, Solomon and O'Neill (1998) have argued that this is a necessary consequence of the nature of mathematics. It is this issue that I intend to explore in this paper, considering the ways in which mathematical texts portray mathematics and the activity of mathematicians.

I have argued elsewhere (Morgan, 1996a) that, although research related to mathematical language has tended to treat this as if it were a single undifferentiated entity, there are in fact different genres of both spoken and written mathematical language appropriate to different social purposes (see also Mousley and Marks, 1991). The notion that particular forms of writing are appropriate for particular purposes involves the employment of the value system of the social context in which the writing plays a role and hence involves the power relations within that social context (Fairclough, 1992). Thus what may be judged appropriate ways of writing for a primary school text book or for a popular book on recreational mathematics may not be judged appropriate for an advanced text book or research paper. Most of us would, however, recognise each of these genres as mathematical – to some extent. But are some more mathematical than others and what might this mean? It is possible to argue that labelling a text as mathematical is a purely conventional matter, depending solely upon the values of those with the power to enforce such a labelling (Morgan, 1998a). A teacher has the power to say a student's text is or is not mathematical; an academic mathematician has the power to judge a school text or curriculum to be or not to be 'really' mathematical (though this power may be contested by other groups, including politicians, parents and pupils as well as teachers themselves). For those of us who engage in doing, reading and writing mathematics at various levels, however, this does not feel entirely satisfactory. We do not experience distinctions between mathematical and not-mathematical as arising from arbitrary whim. So what is it that makes us recognise a mathematical text as being mathematical and are those characteristics *abstract, formal, impersonal* and *symbolic* essential?

WHAT MAKES A MATHEMATICAL TEXT MATHEMATICAL?

The three statements shown in Figure 1 may all be seen to express the same general mathematical fact about the diagonals of rectangles. Yet there are differences in the forms of the statements that affect the ways in which readers respond to them and, in particular, the ways in which readers judge the statements as examples of mathematical writing or as being suitable for use in specific mathematical or educational contexts. In preparing to write this paper, I asked a number of people involved in mathematics education (primary and secondary teachers of mathematics and mathematics education researchers) which of these statements was the 'most mathematical'? The fact that most of those I asked were able to respond to this question without any apparent difficulty suggests that they were able to accept the idea of a hierarchy of forms of writing along a dimension of 'being mathematical'. Most people I asked agreed that it was statement 1 [2]. But what is it that is 'more mathematical' about statement 1? What does it say that is different from what the other statements say?

Statement 1	Statement 2	Statement 3
A rectangle has equal diagonals.	If you measure the lengths of the diagonals of a rectangle, you will find that they are the same.	The measurements of the lengths of the diagonals of a rectangle are always equal.

Figure 1: Three different statements of the 'same' mathematical fact.

There are a number of differences between these three texts. I shall discuss three that seem to relate directly to the nature of the mathematical subject matter:

redundancy: Unlike statement 1, both statements 2 and 3 elaborate the concept of equal diagonals by referring to the equality of the measures of their lengths. We may assume that, within the context of the area of mathematics in which these statements appear to be located, the 'equality' of diagonals must relate to their length. Explicitly stating that the lengths of the diagonals are equal introduces redundant information. This redundancy may lead readers to assign relatively low value to these two texts for several reasons. Firstly, it violates the general principle of communicating only what is necessary and relevant (Grice, 1975) (and hence may be seen as badly written in a general sense). Conciseness and using only that which is necessary and sufficient are highly valued in advanced mathematical discourse; redundancy is thus unlikely to be a desirable feature of mathematical texts [3]. It may suggest to some readers that the author is insufficiently knowledgeable in the field to recognise the existence of this redundant information or that the audience being addressed is one with little mathematical sophistication (hence locating the text in an elementary mathematics discourse, suitable for non-specialists but not for 'real' mathematicians).

temporality: Statement 3 introduces a temporal element into its expression by claiming that the relationship stated is 'always' true. This is a special case of redundancy in that it is a fundamental assumption about mathematical facts that they are true for all time. The general consequences of redundancy were described above. In this case, there is the additional issue that making such a temporal claim may be interpreted as arising from an apparent lack of understanding of the timeless nature of mathematics and the universal nature of mathematical statements. While it might be argued that statement 1 appears to refer only to a single particular rectangle, those familiar with advanced mathematical discourse are more likely to read its form as equivalent to that of the explicitly general statement "If any shape is a rectangle then its diagonals are equal".

human activity: The grammatical subject of statement 2 is the human actor 'you'. The theme of this statement is thus the human activity of measuring and discovering a relationship rather than the relationship itself. But mathematical facts are not dependent on empirical verification. To suggest that the equality of the diagonals is dependent on the result of the human activity of measuring thus again demonstrates a lack of understanding of the nature of mathematics.

Or does it? Is it a purely social phenomenon that we interpret texts in this way, seeing the statements 2 and 3 as examples of immaturity or lack of mathematical understanding? They are, after all, just as true as the first one (though the references to measurement might be challenged on the grounds that measurements are not going to be accurate enough to demonstrate equality) [4]. They can, rather than being seen as less mathematical, be seen as examples of different mathematical genres, expressing different aspects of mathematics for different purposes, for example, to instruct a student or to display the way in which the mathematician discovered the phenomenon. Why should we privilege the formal (non-redundant, timeless, non-human, context-independent) text as more mathematical?

In the analysis above, I have used the idea of conformity or lack of conformity with 'the nature of mathematics' as a means of distinguishing between the texts. The independence of mathematical facts from time and from human action is characteristic of absolutist philosophies of mathematics, which dominate the everyday assumptions of mathematicians and school mathematics (Ernest, 1991). I am employing these notions because they are mainstream assumptions about mathematics and can thus be used to explain common responses to these statements, not because I am committed to such a philosophy of mathematics.

THE RELATIONSHIP BETWEEN FORM AND CONTENT

Although it is possible to say that all three of the statements discussed above are in some sense 'about' the same mathematical relationship, the 'content' of each text is not independent of the form in which it is written. The grammar of the text is part of the system of choices (Halliday, 1985) available to the writer (consciously or unconsciously) to shape the nature of what they are trying to convey; it also forms part of the resources that the reader uses to make sense of the text. The choice between passive and active mood, for example, is not solely a choice about form but is also a choice about whether to represent or to obscure the role of the agent in the process. The grammatical metaphor of nominalisation (using the nominal *rotation* rather than a phrase using the verb *rotate*) is used extensively in mathematical and other scientific texts (Halliday, 1998). It is not simply a different way of expressing the same meaning, but actually allows the writer to say new things such as *this set of rotations forms a group*. The themes with which writers choose to start their sentences affect the logical structure of the text and the ways in which readers may perceive the logic of the subject matter. See Morgan (1996a) for a fuller discussion of relationships of grammar to mathematical meanings.

In the discussion that follows, a small number of texts written by acknowledged mathematicians will be considered, looking in particular at the ways in which the role of human mathematicians in mathematics and mathematical activity is constructed. I shall be pointing to some of the grammatical features of these texts, offering my reading of the ways in which the authors' choices differ and may be seen to construct different pictures of mathematical activity.

MATHEMATICS AS AN AUTONOMOUS SYSTEM *VERSUS* MATHEMATICS AS A HUMAN ACTIVITY

Does mathematics have an autonomous existence or is it the product of humans? Discussions of the philosophy of mathematics generally seem to suppose that these two views of mathematics are separate and incompatible. Yet even the greatest formalist *does* mathematics. A more useful distinction for the current discussion is between mathematical 'facts' and mathematical 'activity'. This allows us to side-step the question of whether the mathematical facts are discovered or invented. The distinction parallels that drawn by Richards between different mathematical discourses: on the one hand he identifies "school math" and "journal math", which focus on the established facts of completed mathematics, making use of "reconstructed logic", while in contrast he identifies "inquiry math" and "research math", which focus on the practice of doing mathematics, making use of the "logic of discovery" (1991, pp. 15-16). The latter two types of discourse are largely oral [5] while the first two are dominated by written texts. The conventional view of formal mathematical writing thus privileges the facts and obscures the activity. In analysing the examples that follow I shall argue that this conventional view is neither truly descriptive of the writing produced by mathematicians nor a necessary consequence of the nature of mathematics.

A historical example

I shall start by considering an extract from a letter published by William Hamilton about his work on quaternions. This extract, together with others from Hamilton's writing on this topic, has been discussed by Solomon and O'Neill (1998) and I shall take their analysis as a starting point in the discussion that follows.

> *My train of thought was of this kind.* Since $\sqrt{-1}$ is in a certain well-known sense a line perpendicular to the line 1, *it seemed natural* that there should be some other imaginary to express a line perpendicular to both the former; and because the rotation from 1 to this also being doubled conducts to -1, it also ought to be a square root of negative unity, though not to be confounded with the former. Calling the old root, as the Germans often do, *i*, and the new one *j*, *I inquired what laws ought to be assumed* for multiplying together $a+ib+jc$ and $x+iy+jz$. *It was natural to assume* the product $= ax - by - cz + i(ay + bx) + j(az + cx) + ij(bz + cy)$; but what are we to do with *ij*?
>
> (Hamilton 1884, cited in Solomon and O'Neill, 1998, p. 214, my italics)

Solomon and O'Neill's analysis of this extract rests on the conventional view of mathematics as separate from the practice of mathematicians. They identify what they describe as "two distinct component texts": a narrative "supertext", characterised by the use of the past tense and temporal ordering (in italics in the extract above), and a "mathematical sub-text", characterised by the "timeless present" and logical ordering, gaining its cohesion from the use of expressions such as *since* and *because*. Their implication is that the narrative of Hamilton's processes of inquiry is not mathematics.

It is certainly possible to distinguish the two parts of the text by these grammatical differences, but is it possible to make a different interpretation? Is the narrative of Hamilton's thinking and feeling any less mathematical than the statements of his results? Hamilton's claims that it was "natural" to infer a second imaginary with particular properties and to construct the product of two hypercomplex numbers in a particular way are essentially mathematical claims, arising from a deep enculturation into mathematical ways of thinking, resting on a conception of mathematics that assumes regularity, connectedness and the legitimacy of using analogy to discover (or construct) new types of object. The claims would not make sense to a reader who did not share this enculturation. I would argue that it is not necessary to separate the two parts of this text; indeed, to do so would be to distort the functioning of the text. This is not a mixing of two texts, one describing Hamilton's activity and the other presenting the mathematical facts. It is a single text that presents a picture of the construction of mathematical knowledge through the human activity of speculating and reasoning – the logic of discovery is human logic as much as it is mathematical logic.

Two present day examples

While historical examples are interesting in themselves and useful in that they can help us to question some of our assumptions, my primary interest lies in looking at the texts that are being constructed by present-day mathematicians and at the writing that students of mathematics encounter and are expected to write. The published work of mathematicians does not now include narratives of discovery like that in Hamilton's letter. However, one clear result that has emerged from the analysis of mathematics research papers that Leone Burton and I have undertaken is that there is not a single way of writing mathematically but that there is enormous diversity (Burton and Morgan, 2000). I will discuss two examples of texts, extracted from published research papers, which construct rather different views of the place of the human mathematician within mathematics. In both cases, the extracts are taken from sections of each paper that introduce the problems to be dealt with, presenting and giving reasons for the notations and procedures to be used. Both start with an announcement of the mathematical topic that "we" will attend to in the coming section. They are also similar in that, unlike Hamilton, both use the present tense and make conventional academic use of *we*.

Extract 1

We shall consider the steady flow of a compressible viscous fluid above an infinite flat disc rotating with an angular velocity Ω about the z-axis. Thus, ... the effects of streamline curvature and Coriolis force must be taken into account. The problem is formulated in terms of cylindrical polar coordinates to take advantage of the axisymmetric motion ... Thus we introduce the non-dimensional coordinates and velocities (r, θ, z) ...

Since the fluid is compressible the variations of the density of the fluid must be considered. Hence, the Navier-Stokes equations and the continuity equation are not sufficient to solve the problem. Thus, we must also have an equation of state and an energy equation.

In this first example, the actions of the mathematician appear to be driven by the logic of the mathematics. The use of *thus* and *hence* in the thematic position at the beginning of sentences highlights the importance of this logic to the development of the text and subordinates the essentially human actions of introducing forms of notation and equations by presenting these as arising from logical necessity rather than from conscious decision making. The role of the mathematician is further obscured by use of the passive. Even the problem itself "is formulated ... to take advantage", without acknowledging the human actions of posing problems and making judgements and decisions in the course of the problem solving.

174

Extract 2

We will now look more closely at the algorithmic aspects of Poincaré's Theorem. We wish to produce a mechanical procedure which ...

What kind of mathematical model of a computing machine is necessary in order to carry out the procedure described in the preceding papers? ... We need to be able to handle real numbers not as sequences of bits but as entities. We need to be able to compare two real numbers ...

Let us go through the steps of the computation to see what kinds of operations are necessary. We need to start by making a decision as to how to represent the input data. ...

In contrast, in this second extract the problems appear to arise from the mathematician's wishes and needs and it is these which drive the development of the logic. The importance of this human motivation is highlighted by the repeated "We need ..." in thematic positions. The "decision as to how to represent the input data" is also seen as a human decision rather than one that is determined solely by mathematical facts and relationships. Although it does not tell a story in the past tense of the author's actions leading up to the conclusions reported later in the paper, this extract does have a temporal aspect: "We will *now* look ...", "We need to *start* ...". It may thus be seen as a narrative of the activity of mathematicians as they read the paper and participate in the logic of discovery [6]. This contrasts with the timeless presentation of extract 1.

While it is possible to identify the different pictures of the activity of human mathematicians presented in these two texts, there are still a number of important questions that need answering. To what extent are these differences functions of the mathematical topics being addressed and to what extent is it a matter of choice for individual authors? Could each text be rewritten in the style of the other? If this were done, what would change? Would it make any difference to the journal editors' decisions about suitability for publication [7]? Knuth (1985) has pointed out differences in the forms of argumentation used in texts about different areas of mathematics. It seems a credible hypothesis that the presentation of human activity in texts about some areas of pure mathematics might also differ considerably from that in papers about statistics or other applied areas. There are certainly differences between different journals in the style of writing in the papers published but it is unclear exactly how this might be related to the topics.

Discussion

As we have seen in extract 2, a temporal aspect is not entirely absent from present day mathematics research publications. Even in those cases where a journal text or advanced textbook conforms more to our stereotypical image of the conventional genre, lacking the temporal narrative of inquiry, it still constructs a role for human beings. In even the most formal and impersonal of texts, the use of imperatives, as Rotman argues (1988), supposes a reader who engages with the text either as a "thinker", joining the author in constructing a mathematical world (by defining, naming, supposing – these functions are also often performed by "we", making the human involvement explicit), or as a "scribbler", performing the calculations that are necessary to complete the argument. Such texts tend to use what Richards (1991) calls the logic of reconstruction rather than the logic of discovery. Yet, as Livingston argues, even the most formal proof contains within itself the actions of the human mathematician who constructed it:

> ... the rigor of a mathematical proof not only lies within, but is hidden within, the proof's lived-work, ... a proof consists of the pair the-material-proof / the-practices-of-proving-to-which-the-proof-is-irremediably-tied, and ... a proof is cultivated so as to realize the material proof as a disengaged version, or account, of that proof's lived-work.
> (Livingston, 1986, pp. 176-177)

Solomon and O'Neill claim that "mathematics cannot be narrative for it is structured around logical and not temporal relations" (1998, p. 217). As soon as you admit that mathematicians *do* mathematics, however, the doing must take place in time. The product

$$(a + ib + jc)(x + iy + jz)$$

did not equal

$$ax - by - cz + i(ay + bx) + j(az + cx) + ij(bz + cy)$$

until Hamilton made his assumption that this was the way in which such products should be formed. It is only after and through the "lived work" of the mathematician that the timeless truth of the self-consistent logical system emerges. Rather than saying that it is only the timeless logic and the statement of eternal facts that is truly mathematical while Hamilton's personal narrative is extra-mathematical, it is possible to argue that presenting mathematics as an eternal and autonomous system actually misrepresents the nature of mathematics as the product of human activity.

MATHEMATICAL WRITING IN SCHOOL

My concern in this paper is not purely with the nature and diversity of texts produced by mathematicians, but with the implications that this may have for students of mathematics and their experiences of mathematical writing. There are, of course, many differences between the texts of professional mathematicians and those produced by students of mathematics in school [8], not only in the topics addressed but also in the forms the writing takes. One reason for this is obviously because of differences in rhetorical function. The writing that school students do in mathematics classes generally serves or is intended to serve two kinds of functions. On the one hand, it is seen as part of the learning process (assisting reflection or recording for future reference) and, as such, is private – written by the student for herself (Connolly and Vilardi, 1989; Emig, 1983). On the other hand (and sometimes at the same time) it is an important medium through which evaluation takes place – and the writing is thus addressed to a teacher or examiner. There is an interesting parallel with the writing of professional mathematicians here: for many mathematicians, much of their mathematical activity takes place through/in writing and the product of their activity is also judged through its presentation in writing. In the case of the mathematicians, it is clear that the two functions are served by very different genres of text – nobody would consider evaluating mathematical research on the basis of the mathematician's rough workings. In school this distinction is not so clear cut.

In the present mathematics curriculum in England and Wales there are (at least) two very different genres in which students must learn to write. These relate broadly to distinctions between content and process, between facts and activity, between reconstructed logic and the logic of discovery. The first genre (and the one that dominates school mathematics) is that described by Ernest as follows:

> During most of their mathematics learning career from 5 to 16 years and beyond, students work on textual or symbolically presented tasks. They carry these out, in the main, by writing a sequence of texts (including figures, literal and symbolic inscriptions, etc.), ultimately arriving, if successful, at a terminal text, "the answer".   (1998, p. 255)

As Ernest demonstrated in an example presented in an earlier paper (Ernest, 1993), the textual sequence presented for evaluation does not necessarily match the sequence of the student's activity but is a reconstruction that conforms to the logic of the format required by teacher and examiner. As teachers of simple linear equations know, the injunction to 'show your working' is intended to coerce students to produce texts demonstrating that they can 'do the same thing to both sides' rather than to record their informal processes of inspection or trial and error. Learning the rules of this genre appears to be unproblematic for most pupils – they can generally produce texts in the required form even if the sequence of statements does not logically and successfully lead to a correct answer.

The second major school mathematics genre is explicitly about processes. The National Curriculum for England and Wales (Department for Education, 1995) expects students to be taught processes such as breaking complex problems into manageable tasks, making generalisations, and explaining their solutions [9]. One of the main ways that achievement in these areas is assessed is through students' written reports of their work on investigative or other substantial problem solving tasks.

The texts that students produce for GCSE coursework [10], for example, are expected to contain some form of narrative; a text written in an impersonal style may well be considered inappropriate by teacher-assessors (Morgan, 1996b). My study of students' reports of their investigative work (Morgan, 1995; 1998b) suggests that at least some students, while attempting to follow their teacher's instruction to "write down what you did", failed to do so in a way that might be considered mathematical. Some included aspects of their work that were deemed inappropriate by teacher-assessors, including mention of social aspects of the work, meta-cognitive reflection or affective responses. One student's comment that "with such a definite pattern a formula should be easy to find" was greeted with laughter by teachers and may have contributed to their low evaluation of his intellectual competence. A narrative by another student (see Figure 2), attempting to tell the story of her investigative activity, was greeted with hostile parody from one teacher:

> That's one of my pet hates. It's the sort of, you know, "Miss put this task on the board and we copied it down" but not quite that bad.

while teachers generally responded negatively to use of the unelaborated statement "I found a formula". Students who produced such texts had failed to distinguish between the parts of their activity that would be judged to be mathematical and those parts their teachers would see as not-mathematics.

Once Suzanne and I had completed tasks 1 and 2, we set out to discover if there was any connection between the triangular area and the lengths of the sides. First we lettered the sides:-

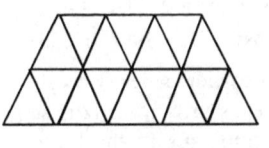

a = slant line
b = base line
c = top line
d = area

In a very short time we had discovered a relationship between the lengths of the sides and the area (triangular). We were able to put this into a formula:-

$$ab + ac = d$$

Figure 2: An example of a student's 'inappropriate' narrative

We cannot assume that students will 'pick up' the desired characteristics of mathematical writing implicitly. Indeed, the less explicit the 'rules' of mathematical writing are, the more likely it is that students from working class backgrounds and from other non-dominant social groups will be disadvantaged. Such students are likely to lack the linguistic skills and cultural capital that would enable them to recognise the forms that are valued within the particular discourse and to realise these themselves in order to construct 'appropriate' mathematical narratives (Bernstein, 1996; Bourdieu and Passeron, 1990).

It seems that students are much less successful at learning the rules of the investigation report genre than the traditional 'steps leading to an answer' genre. I would suggest that there are several possible reasons for this:

- Far less time in mathematics classes is spent working with investigation reports than with traditional texts.

- Most mathematics teachers' personal experience of mathematical writing during their own education was heavily dominated by the traditional genre.

- Students seldom read examples of investigation reports that have been written by their peers – and even less frequently read examples of this genre written by 'experts'.

- Mathematics teachers may not see it as part of their role to teach their students how to write and, even if they do undertake this, they lack a language with which to identify and describe the features of the genre of investigative reports (Morgan, 1998b).

The first of these reasons is probably unavoidable given current curriculum constraints, while the second is unlikely to change much in the near future. The last two reasons, however, are more susceptible to the effects of teacher education and curriculum development within existing structures.

CONCLUDING REMARKS

The analysis that I have offered above of the representation of mathematical activity in texts produced by research mathematicians may suggest one way forward. Hamilton's narrative of his train of thought, the logic driving the mathematician's activity in Extract 1, the questioning and expression of needs and wishes in Extract 2 all provide their readers with (different) pictures of the processes as well as the products of mathematical activity. Recognising that such alternatives exist and that mathematical writing does not have to be abstract, formal and impersonal may, in itself, help to make mathematical writing and mathematical activity more accessible to some of those students who have previously felt excluded by the apparent lack of human presence in mathematics.

The features that I have identified in forming the analysis may provide a step towards a language of description of aspects of mathematical text that could be used by teachers to identify desirable characteristics and communicate with students about them (see also Morgan, 1996a). Perhaps these and other examples could also provide more appropriate models for teachers to use to support student writing than the traditional style they are more familiar with from their own induction into mathematics. Some of the forms of language used to introduce the representation of the mathematicians' processes (*I inquired ...*; *Thus we introduce ...*; *We need to be able to ...*) might, with adaptation and augmentation by forms useful for representing other aspects of problem solving activity (including those that have already been found to be acceptable within the genre), provide a basis for producing mathematical writing frames (Lewis and Wray, undated) to support student writing – a strategy that has already been shown to have some potential to help students produce acceptable forms of investigation report (Lee, 1997).

This is just one possible approach to supporting student writing. More generally, mathematics educators need to generate more explicit knowledge about the language that may be used to communicate about mathematical activity. Having knowledge of what forms of writing may be considered 'appropriate' and 'mathematical' – and what alternative forms of writing are available – is a necessary prerequisite for supporting student writing. The analysis that I have presented here of ways in which mathematical activity may be presented provides a step towards generating this knowledge and I have suggested a way in which it might be made accessible to teachers and students through the practical classroom tool of writing frames. There are clearly many other aspects of mathematical writing yet to be analysed. Moreover, when further knowledge about mathematical language has been generated it will still need to be transformed into forms that are usable in the classroom and into teaching and learning tasks. It is this process that will give more students the chance to learn to write effectively about their mathematical activity.

NOTES

1. Attempts to change the conventions of academic writing have often been inspired by feminists, challenging male domination of the academy by challenging its pretension to impersonal neutrality. See, for example, (Cameron, 1992).

2. Those who did not give this answer generally turned out to have been addressing a rather different question, for example: Which do you prefer? or Which is the easiest for children to understand?

3. On the other hand, Kane's (1968) investigation of mathematical English in school textbooks identified some examples of high levels of redundancy.

4. Within the current UK school context, measurement is often used as a means of 'establishing' such mathematical facts and relationships. It seems likely, therefore, that school students and, perhaps, their teachers would be likely to produce statements like statements 2 and 3. This empirical approach is, however, in conflict with the deductive reasoning that is characteristic of mathematical proof and of the formal presentation of most higher mathematics texts. The criticism by university mathematicians of current undergraduates' lack of appreciation of the nature of proof (see, for example, London Mathematical Society, 1995) may be exacerbated by the patterns of language that students have established in school. The types of writing that school students engage in, by focusing on empirical 'results' and student activity, do not help them to establish high status abstract forms of mathematical discourse – and so lead to their condemnation as empiricists.

5. An obvious exception to this in the UK school context is the writing of reports of investigative work in which students are expected to provide a narrative of their problem solving activity (Morgan, 1998b). Some forms of recreational mathematics may also provide examples of written genres within the inquiry math discourse.

6. Of course, this logic may not follow the same sequence as the work done by the author to 'discover' the original results. The paper is a reconstruction of that work, not a recount.

7. It is possible that well established and respected mathematicians are more likely than their more junior colleagues to write (or more likely to be published) in less conventional and more personal styles. There is some evidence to suggest that women mathematicians (often insecure in their employment and status) are less likely to depart from the conventional impersonal style (Burton and Morgan, 2000).

8. I will only consider the writing done by students here because it represents the product of their mathematical activity, just as research papers represent the product of the activity of professional mathematicians (though the parallel is by no means exact). Similar questions could also be addressed for other types of school mathematics texts, including textbooks and teacher-produced writing.

9. The National Curriculum has recently been reviewed and the revised version (Department for Education and Employment, 1999) came into effect from August 2000. This still includes reference to processes, though in a different format.

10. GCSE (General Certificate of Secondary Education) is the examination taken by almost all students in England and Wales at age 16+. Most syllabuses include a coursework component that is assessed by students' own teachers and moderated by an external examination board. This generally consists of reports of one or more investigative or problem solving tasks.

REFERENCES

Bernstein, B.: 1996, *Pedagogy, Symbolic Control and Identity: Theory, Research, Critique*. London: Taylor & Francis.

Bourdieu, P. and Passeron, J.-C. 1990, *Reproduction in Education, Society and Culture* (2nd ed.). London: Sage.

Burton, L. and Morgan, C.: 2000, 'Mathematicians writing.' *Journal for Research in Mathematics Education, 31*(4).

Cameron, D.: 1992, *Feminism and Linguistic Theory*. (2nd ed.). Basingstoke: Macmillan.

Connolly, P. and Vilardi, T. (eds.):1989, *Writing to Learn Mathematics and Science*. New York: Teachers College Press.

Department for Education: 1995, *Mathematics in the National Curriculum*. London: HMSO.

Department for Education and Employment: 1999, *Mathematics: The National Curriculum for England*. London: DfEE.

Emig, J.: 1983, 'Writing as a mode of learning.' In D. Goswami and M. Butler (eds.), *The Web of Meaning: Essays on Writing, Teaching, Learning, and Thinking* (pp. 123-131). Upper Montclair, NJ: Boynton/Cook Publishers.

Ernest, P.: 1991, *The Philosophy of Mathematics Education*. London: Falmer Press.

Ernest, P.: 1993, 'Mathematical activity and rhetoric: A social constructivist account.' In N. Nohda (ed.), *Proceedings of the Seventeenth International Conference for the Psychology of Mathematics Education* Vol. 2 (pp. 238-245). Tsukuba, Japan: University of Tsukuba.

Ernest, P.: 1998, 'The culture of the mathematics classroom and the relations between personal and public knowledge: An epistemological perspective.' In F. Seeger, J. Voigt, and U. Waschescio (eds.), *The Culture of the Mathematics Classroom* (pp. 245-268). Cambridge: Cambridge University Press.

Fairclough, N.: 1992, 'The appropriacy of "appropriateness".' In N. Fairclough (ed.), *Critical Language Awareness* (pp. 33-56). Harlow: Longman.

Grice, H. P.: 1975, 'Logic and Conversation.' In P. Cole and J. L. Morgan (eds.), *Syntax and Semantics* Vol. 3 (pp. 41-58). London: Academic Press.

Halliday, M. A. K.: 1985, *An Introduction to Functional Grammar*. London: Edward Arnold.

Halliday, M. A. K.: 1998, 'Things and relations: Regrammaticising experience as technical knowledge.' In J. R. Martin and R. Veel (eds.), *Reading Science: Critical and Functional Perspectives on Discourse of Science* (pp. 186-235). London: Routledge.

Kane, R. B.: 1968, 'The readability of mathematical English.' *Journal of Research in Science Teaching, 5*, 296-298.

Knuth, D. E.: 1985, 'Algorithmic thinking and mathematical thinking.' *American Mathematical Monthly, 92*, 170-181.

Lee, C.: 1997, *The Use of Teaching Strategies to Improve Students' Writing of Mathematics.* Unpublished Report to the Teacher Training Agency.

Lewis, M. and Wray, D.: undated, *Writing Frames: Scaffolding Children's Non-Fiction Writing in a Range of Genres.* Exeter: The Exeter Extending Literacy Project.

Livingston, E.: 1986, *The Ethnomethodological Foundations of Mathematics.* London: Routledge & Kegan Paul.

London Mathematical Society: 1995, *Tackling the Mathematics Problem.* London: London Mathematical Society.

Morgan, C.: 1995, *An Analysis of the Discourse of Written Reports of Investigative Work in GCSE Mathematics.* Unpublished PhD dissertation, Institute of Education, University of London.

Morgan, C.: 1996a, '"The language of mathematics": Towards a critical analysis of mathematical text.' *For the Learning of Mathematics, 16*(3), 2-10 + erratum *17*(1).

Morgan, C.: 1996b, 'Teacher as examiner: The case of mathematics coursework.' *Assessment in Education, 3*(3), 353-375.

Morgan, C.: 1998a, 'Assessment of mathematical behaviour: A social perspective.' In P. Gates (ed.), *Proceedings of the First International Mathematics Education and Society Conference* (pp. 277-283). Nottingham: Centre for the Study of Mathematics Education, Nottingham University.

Morgan, C.: 1998b, *Writing Mathematically: The Discourse of Investigation.* London: Falmer.

Mousley, J. and Marks, G.: 1991, *Discourses in Mathematics.* Geelong: Deakin University.

Richards, J.: 1991, 'Mathematical discussions.' In E. von Glasersfeld (ed.), *Radical Constructivism in Mathematics Education* (pp. 13-51). Dordrecht: Kluwer Academic Publishers.

Rotman, B.: 1988, 'Towards a semiotics of mathematics.' *Semiotica, 72*(1/2), 1-35.

Solomon, Y. and O'Neill, J.: 1998, 'Mathematics and narrative.' *Language and Education, 12*(3), 210-221

11

BEING CREATIVE WITH THE TRUTH? SELF-EXPRESSION AND ORIGINALITY IN PUPILS' MATHEMATICS

Peter Huckstep and Tim Rowland

University of Cambridge

Whenever one justification for learning mathematics is questioned, educationists have usually been ready to offer an alternative rationale. Often this involves presenting (or re-presenting) the subject in a new light. This paper will consider the implications of trying to invoke creativity in mathematics. We shall argue that 'creativity' is a complex notion and should only be applied to mathematics with caution and attention to meaning.

INTRODUCTION

A recent publication on the aims and purposes of school mathematics (White and Bramall, 2000) has unexpectedly caused something of a furore, attracting some fairly harsh responses in the media. It is not the question *Why Learn Maths?* in the book's title that has generated such alarm and, at times, contempt, but the corollary that concerns the extent to which mathematics should remain a compulsory subject within the years of compulsory schooling. To maintain the status quo, one journalist has mounted a rescue operation. In a *Financial Times* article, Michael Prowse (2000) confidently attempts to provide his own rationale:

> One of the great merits of maths is it teaches the meaning of objective knowledge. Unlike in history or sociology, there is no scope for diverging opinions, and no possibility of disguising one's ignorance. No subject demands more precise reasoning.

But the authors of *Why Learn Maths?* had already explored the limitations of this classic defence. Bramall, for instance, whilst agreeing that "...mathematics ... can supply us with precise, objective and demonstrably correct answers" concludes that

> in so far as mathematics is concerned only with describing and communicating about the quantitative aspects of phenomena, it can be no more than a means to an end (2000, p. 54).

Since, as he continues, "no amount of mathematical calculation will produce a decision about the ends of human life" he argues that subjects such as the social sciences (including history and sociology) – i.e. those areas of the curriculum that *do* address questions of values – can claim educational priority over mathematics. Elsewhere in the book, Huckstep (2000) also tackles the "precise reasoning" argument head-on. He concludes that the objective processes of mathematics make it an ideal vehicle for mental training of a certain kind, but that its scope in this respect is more modest than is commonly supposed.

Yet if the well-established mathematical features of objectivity and precision are insufficient to convince some educators of the unquestionable value of sustained learning of mathematics in an individual's education, perhaps we can find some justification by considering aspects of those less exacting, more subjective areas of the curriculum. For example, we might want to argue that mathematics is a source of *creativity*.

Some mention of the potential for developing 'creative capabilities' in mathematics is made by Ernest in his contribution to *Why Learn Maths?* but stronger support can be found in the preamble to the latest revision of the school mathematics curriculum in England which includes the following passage:

> Mathematics is a *creative* discipline. It can stimulate moments of pleasure and wonder when a pupil solves a problem for the first time, discovers a more elegant solution to that problem, or suddenly sees hidden connections. (QCA, 1999, p. 36, our emphasis)

The assumed symbiosis, here, of the notions of creativity and discovery in mathematics is notable. As we shall demonstrate, creativity is often invoked in order to draw out a contrast with discovery. 'Discovery', in turn, is inevitably intertwined with a Platonic view of mathematical knowledge.

The QCA manifesto for creativity as a dimension of school mathematics is certainly welcome. The relevant paragraph can only hint at what 'creative' might mean in a mathematical context, and an overwhelming reaction might well be to assent to it without feeling the need to ponder its meaning. Closer inspection suggests that, in the passage quoted above, this has something to do with the *intellectual* action of the pupil as a stimulus and cause of experiences of an *emotional* kind.

Craft (2000) in her thorough exploration *Creativity Across the Primary Curriculum,* admits that mathematics (and science) "...sometimes get ignored in discussions of creativity" (p. 77) and offers some suggestions of when the subject might involve creativity. A more diverse view of creative action and experience in mathematics teaching and learning is taken by the authors of *Creative Mathematics: Exploring Children's Understanding* (Upitis, Phillips and Higginson, 1997). We have found both books to be a useful source against which to test and compare some of the perspectives on creativity that we find elsewhere.

ON 'CREATIVE' AND CREATIVITY

'Creative' is a classic example of what philosophers have called a 'hooray' word. In this respect, it sits alongside 'rational', 'freedom', 'culture', 'democracy' and 'art'. The meanings of hooray words are typically imprecise; they often have contestable definitions, yet all are heavily value-laden. Such words can have considerable rhetorical force in assertions: to suggest that an educational policy, a subject on the curriculum or a teaching approach promotes or involves creativity is usually sufficient to commend it. It is usually not thought to be necessary to know precisely what is meant by 'creativity'; it is simply assumed to be an unassailably Good Thing, like motherhood and apple pie. Indeed, since God is the ultimate creator, one may suppose that creativity, along with cleanliness, is next to godliness.

During the 1960s and the 1970s, philosophers of education made several sustained attempts to analyse the concept 'creative', and subsequently several chapters and articles on the topic appeared in the literature of the theory of education (Dearden, 1968; Elliott, 1971; White, 1972; Wilson, 1977). Teacher-training courses around that period often treated creativity as one of the key aims of education.

Favoured aims in education ebb and flow in their prominence, yet the attraction of creativity in education has by no means vanished. But what is meant by creativity? The *Oxford Companion to The Mind* (Gregory, 1987) points out that it is one of those terms (like 'intelligence') that "...direct us in practice to a number of concerns that are rather separate".

> Some of these, like 'innovation' and 'discovery' have a bearing on *the ideas or objects that people produce*; some like 'self-actualization' refer more to *the quality of life an individual leads*; and some like 'imagination' and 'fantasy', point us in the first instance, to *what goes on in a person's head*. (p.171 our emphasis) [1]

The *Companion* immediately alerts us to the portmanteau character of the word, a kind of catch-all for a range of behaviours or qualities – all somehow made instantly commendable by association with the C-word. Any examination of creativity in mathematics must establish which of these concerns – (i) ideas/objects (ii) quality of life or (iii) thought processes – (if any) is appropriate, or whether there is something unique to mathematics that prompts us to consider conceptions in addition to those raised in the *Companion*.

MATHEMATICS, MATHEMATICIANS AND CREATIVITY

Creativity has *traditionally* had its home in the arts. Elliott (1971) has argued that "the myth of divine creation" runs deep in our traditional idea of 'creative' and for this reason it is the artist – one who literally *makes* something – who is truly creative. The objects that such arts produce are typically paintings, sculpture, forms of literature and music. Elliot's distinction has echoes as far back as Aristotle who in his *Ethics* distinguished art from science. " The business of every art", he wrote, "is to bring something into existence..." But "scientific knowledge", he insists, " is of things that are never other than they are..." (Aristotle 1955 edn, pp. 174-5). The distinction between art and science (and mathematics, for that matter), even if it can not be made quite as Aristotle does, is one that continues to be maintained.

Elliott claims that notions of creativity which lie outside the arts constitute a second, different version of creativity, one that is concerned with imaginative thinking – the discovery of novel solutions and problem-solving rather that the 'making' of something. The *Cockcroft Report* (DES, 1982) famously asserted that "the ability to solve problems is at the heart of mathematics" (p. 73), and we may suppose that a different version of 'creative' does apply in mathematics. This does not entail the *making* of something, but depends upon the producing of novel ideas, imaginative thinking and problem-solving.

Craft's (2000) account of creativity follows Elliott in this latter respect. She makes use of the notion of imaginative thinking but lays further stress on the central idea of what she calls 'possibility thinking'. Additionally, she incorporates considerations of 'self-actualisation'. But whether this rich account is really exemplified in mathematics is another question. Her mathematical examples are restricted to three main areas: open-endedness, the development of *confidence* by a tolerant and encouraging approach to conjecturing and creative *teaching* of mathematics. Craft makes scarcely any reference to Elliott's first version of creativity which, with its links to the arts, seems to offer more scope for developing the self. In a moment we shall pursue this 'traditional' version a little further.

The view that mathematical entities exist independently of both the mind and notation dies hard. Such Platonism implies that mathematics is something already existing 'out there' which is discovered, not originated, by mathematicians. In a recent study, Leone Burton (1999) interviewed 70 research mathematicians about their beliefs and working practices. She describes how research mathematicians perceive themselves as contributing to the construction of a 'Big Picture', an interconnected unifying story. On the other hand, the Big Picture presented problems of contradiction for some of Burton's participants, like the one who said:

> As mathematicians we want to believe that it is objective but I guess rationally I think it is socio-cultural and emotionally I want to believe it is objective (p.139)

Notwithstanding the fact that research mathematicians – certainly those in Burton's sample – typically work in a cooperative manner and many acknowledge the socio-cultural embeddedness of their work within various communities of practice, Burton concludes that:

> Mathematicians do mostly subscribe to an absolute view of mathematics and what I have called the Big Picture lies not far away from where they position themselves (p.140)

This position is spelled out towards the end of a recent radio programme [2], when interviewer Melvyn Bragg asked mathematician Ian Stewart:

> Is Platonism alive in mathematics? Are mathematicians still looking for what is transcendent?

To which Stewart replied:

> This is a long-running undercurrent and there is something to it. If you talk to professional mathematicians, on the one hand they know that what they do is a collective human activity, and on the other hand they are all closet Platonists. I am! While you're trying to do creative mathematics, while you're trying to – "is it invent, is it discover, I don't know, but I'm trying to do it" … something new, trying to solve a problem that's not been solved before – you have the very strong feeling that the answer, there is only one answer and it's sort of sitting there, and it's your job to find where … what it is. You

can't make it up as you go along. You can make up your exploratory route, and we all do, but you have this mental image of moving through some sort of landscape or some sort of world *looking* for things. And in order to be able to carry out that process, um, it makes it much more possible to do it if you have this illusion that these things exist. How can I look for something that isn't there? [...] So Platonism is alive and well in the operating working philosophy of mathematicians.

A different response to Bragg's question might be offered by those mathematicians and writers on mathematics who try to persuade us that mathematics shares common features with the fine arts, indeed that they are creative in Elliott's traditional sense. On the face of it mathematicians cannot produce the same kind of objects as the artists. For some mathematicians, however, mathematics is not simply discovered. It is, to some extent at least, created, and it is this creative activity which they believe connects artist and mathematician. Usually, it is a similarity between the *thought processes* of mathematician and artist to which attention is drawn. For example, Bocher writes:

I like to look at mathematics almost more as an art than as a science; for the activity of the mathematician, constantly *creating* as he is, guided though not controlled by the external world of the senses, bears a resemblance, not fanciful I believe but real, to the activity of an artist, of a painter let us say. (Moritz, 1914, p. 182, our emphasis)

For Bocher, as for many others [3], it is creative activity which he believes connects artist and mathematician.

G. H. Hardy, on the other hand, goes rather further that this by asserting that the very *objects* that both artist and mathematician create are essentially of the same kind:

A mathematician, like a painter or a poet, is a *maker* of patterns. If his patterns are more permanent than theirs, it is because they are made with ideas. (Hardy, 1967, pp. 84-85, our emphasis)

Pattern is, of course, intrinsic to mathematics. Indeed, for some mathematicians, the subject simply is "the classification and study of all possible patterns" (Sawyer, 1955, p. 12). Yet, whilst a painter, composer, sculptor, poet or choreographer may be typically engaged in 'working out' patterns in producing fine (or expressive) art, it is not clear that a mathematician 'works out' pattern in the same way. We need to distinguish between 'working out' in the sense of *originating* a pattern from scratch, as it were, and 'working out' in the sense of identifying and describing an already existing pattern. Higginson, in *Creative Mathematics* (Upitis *et al.*, 1997) endorses this distinction when he remarks that:

Human beings are pattern-*seeking* and pattern-*creating* creatures. Tessellations and tiling patterns have a long and a culturally rich history. They can be seen as *visual poetry*. (p. 50, our emphasis).

187

To associate mathematics with poetry in this way is surely a misleading metaphor. A poet can always create patterns, but it is questionable whether a mathematician can do any more than seek (and explain) them.

Indeed, an insightful remark from Popper suggests that in creating *some* aspects of mathematics others must necessarily be discovered:

> The series of natural numbers which we construct creates prime numbers – which we *discover* – and these in turn create problems of which we never dreamt. (1972 p. 138) [4]

The sequence of prime numbers certainly presents a plethora of complex problems, and it may *seem* as though both the primes *and* the problems have been thereby created rather than discovered. The same may be said of pattern. Even if we admit that mathematics is created – or perhaps, more neutrally, that it is constructed – it does not follow that the particular patterns which are discovered later had ever been anticipated or, in Popper's terms, that mathematicians had ever dreamt of them.

Leaving aside the specific issue of creating pattern, it is worth pausing to consider a more general objection from Elliott that the 'traditional' concept of creativity can have ironic undertones when it is put to use in some contexts. Elliott writes:

> If we call someone a 'creative' historian it is virtually impossible, no matter how we load the context, to avoid the suggestion that he makes up his stories instead of deriving them from the historical evidence ... 'Creative biologist' suggests a breeder of new germs; 'creative anatomist' a Dr. Frankenstein; 'creative chemist' an alchemist. (1971, p. 140)

Similarly, we might with some justification wonder whether a similar irony is attached to the idea of 'creative mathematics', as it is to 'creative accountancy' where the suggestion is that truth (more strictly, validity) has been displaced in favour of less dispassionate goals [5]. Nevertheless, Elliott does point out that, with science at least, not all applications of the traditional concept of creative are ambiguous. There is at least one class of notable exceptions. This is because great artists do not simply make *things*, we often regard them as having made a *world*. This accomplishment, Elliott believes, can be attributed to great scientists too, those who " ... have quite radically re-structured our world, which is the world as we conceive – and even perceive – it." (1971, p. 144)

Alongside the revolutionary scientists whom Elliott cites, it would seem reasonable to include certain mathematicians who, in the same way, we could intelligibly regard as being creative. Amongst these we could include, for example, mathematicians such as Gauss, Bolyai and Lobachevsy, who have shown that new and different geometries are possible, that Euclidean geometry does not finally and conclusively describe the space of the world in which we live, as was once unshakeably believed.

One writer who has argued that precisely this kind of world-making constitutes a development of the *self* is Sullivan who wrote:

> ...it is certain that the real function of art is to increase our self-consciousness; to make us more aware of what we are, and therefore of what the universe in which we live really

is. And since mathematics, in its own way, also performs this function, it is not only aesthetically charming but profoundly significant. It is an art, and a great art. It is on this, besides its usefulness in practical life, that its claim to esteem must be based. (1960, p.2021)

But to link mathematics and art to the self by reference merely to *consciousness* is a fairly weak claim when we consider how accounts of the arts so often invoke the more dynamic notion of self-*expression*. Indeed, Dearden in his classic *The Philosophy of Primary Education* offers 'self-expression' as part of his analysis of creativity. His is, however, something of a mechanistic model in which the emotions are somehow "corked up", and creativity is the process of releasing tension caused by this suppression. This can take on a variety of forms, some recreational, including activities in which one is literally both creative and re-created, for example by fashioning a beautiful garden, or engaging in music-making, painting and so on. The more symbolic expression of grief within the rituals of a funeral also captures some of what Dearden seems to mean by this sense of creativity. Certainly, mathematical activity – puzzles and 'recreations', for example – can be a diversion from 'reality', which often looks much messier and less controllable. Equally, the tidy mathematical solution that emerges from a chaotic stack of scribbled false starts and dead ends has a cathartic function, releasing an internal mêlée of some kind.

Perhaps it is not too far-fetched to link this emotional dimension, or outcome, of mathematical activity with the self-actualisation which Craft (2000) includes in her account of creativity. One of Leone Burton's questions to the research mathematicians regarding their production of new knowledge was along the lines 'How do you know when you know?'. Burton quotes one mathematician who replied:

> When I think I know, I feel quite euphoric. So I go out and enjoy the happiness. Without going back and thinking about whether it was right or not but enjoy the happiness. When I discover something, I just enjoy the feeling. It is almost pointless to try and check it because you are so euphoric that you cannot possibly check (Burton, 1999, p. 135)

Burton comments that *feelings* such as these permeate the mathematicians descriptions of coming to know, and that such an experience "holds them in mathematics". Burton's choice of words here is interesting; perhaps it is not too fanciful to propose that they suggest an embrace. She continues:

> Whether your knowing is robust, or not, for the moment that you know that you know the power of that knowledge lies in the feelings it evokes not externally in the mathematics (p.135).

But even if mathematics were on occasions a vehicle for such a release, it would surely require more argument to show that it is in this respect that much of its curriculum importance lies [6].The time has come for us to proceed from the symposium into the schoolroom.

LEARNING AND CREATIVITY

We begin by asking whether we can really treat pupils in school as though each one is something of a young world-making Gauss in the making. Upitis in *Creative Mathematics* takes her own enthusiastic comparison of the arts and mathematics to excess, when she writes [7]:

> As educators, we would like children to learn about spelling and rules of grammar by becoming writers; we would like children to learn about music theory and performance by being composers; we would like children to learn about arithmetic and concepts of mathematics by becoming mathematicians – *makers* of mathematics. (Upitis *et al.*, 1997, p. 36 our emphasis)

There is a modest sense in which this might be true. Ernest suggests that pupils develop what he calls 'creative capabilities' when they are encouraged not only to solve problems but also to '... *pose* mathematical questions, puzzles and problems ...' (2000, p.11 our emphasis). Craft too, lays great emphasis on the act of *conjecturing* in mathematics (2000, pp. 78-79). Certainly the *making* of conjectures is one important task of a fully-fledged mathematician, but Craft draws attention to pupils' feelings of exposure when they make conjectures publicly, in the classroom. In connection with the management of such risks, Rowland (1999) has introduced a metaphysical construct which he names the zone of conjectural neutrality (ZCN). Rowland describes the ZCN as a space between what we believe and what we are willing to assert. The issue central to the notion of ZCN is summarised in the question "Where are pupils' conjectures located? Who is responsible for them?" The default position must be that a conjecture belongs to the one who utters it. If the conjecture is asserted with conviction, better still, if it is subsequently validated as true, then this is not an affective problem. But if a conjecture is offered tentatively, then it is better that it be located somewhere neutral before it is tested, in order that there be some prospect of "testing on a cognitive rather than an affective level" (Dawson, 1991, p. 197). This is in defiance of the cultural norm that the pupil is judged to be 'right' or 'wrong' rather than the 'answer' judged 'true' or 'false'; that it is s/he who is on trial, not her/his beliefs. Rowland draws attention to ways in which pupils use vague language in order to distance themselves from tentative conjectures even as they articulate them. He indicates how teachers can learn to 'read' the modality in pupils' attitudes to their conjectures, and suggests a number of ways in which the teacher can manage the classroom so as to facilitate the placement of a tentative proposal in the ZCN. One way, for example, might be to write it on a flipchart, and say something like "OK, let's take a look at this conjecture", possibly without reference to the one who proposed it or constant application to him/her for arbitration or interpretation.

We can agree, with Craft, that conjecturing activity is creative in the sense of *making* sense (and we return to this aspect of creativity in a moment). A conjecture (other than a pure guess) is the outcome of some unifying cognitive effort, giving rise to a prediction to a situation outside past experience or perhaps a generalisation that makes sense of a range of discrete data. Perhaps the sensitive management of such conjectures is an example of creative teaching? We might be content to call it 'sensitive' or even 'professional' in the case of the teacher who has been sensitised to the significance of vagueness in mathematics talk.

But we struggle to discern whether Upitis' vision – the wish for children to learn mathematics by becoming makers of mathematics – is rooted in any epistemological position or empirical evidence known to mathematics education, or whether it is the 'mere' expression of a seductive kind of optimism, a romantic view of the world in general and childhood in particular. On the other hand, this stunning claim that pupils can actually make mathematics for themselves becomes more tangible, more believable, when we consider what Phillips has to say:

> Part of the making of mathematics involves *sense*-making ... another draws on the application of these concepts to real-world situations and real-world problems ... (Upitis *et al.*, 1997, pp. 143-144, our emphasis)

'Making sense' is a familiar metaphor, and a compelling one. To make sense of something is commonly taken to mean to *understand* it, but that merely shifts the semantic question onto 'understanding'. What is it that we 'make' when we make sense? Is this at all the same kind of making that Elliott describes in his 'traditional' version of creativity, or is it merely some kind of 'doing'? Perhaps there is something to be learned from the French verb *faire*, meaning at once both to make and to do. Thus *faire le ménage* is simply to do housework, and *faire la vaisselle* to do the washing up. Nothing very creative here, it's just getting on with the job. Perhaps, in these examples, 'making' amends for someone who 'made' a mess? Similarly, in German, *Was sollen wir machen?* – what shall we do? Like 'makin' whoopee', there is no pretence to creating something, it's just something that you do.

So there is nothing puzzling in speaking of pupils as makers of mathematics if this is simply elliptical reference to their making sense of mathematics or making applications of mathematics to the extra-mathematical world. One of the central tenets of radical constructivism is the view that knowledge of the world in general (and mathematics in particular) is, quite literally, what any individual makes of it. Mathematical meaning is simply a way of describing our sense-making. Nothing more, certainly nothing less. One reason for learning mathematics is that one can make something *of* it. However, some, if not all, of the applications made follow *from* the learning of mathematics, or arise *in* learning it. Yet in the following rather pithy remark from Upitis there seems to be more than this at stake:

> At the moment, manipulative materials are used primarily to *learn about* mathematics, rather than to *make* mathematics (Upitis *et al.*, 1997, p. 131).

Clearly, to suggest that children are making the mathematics itself, rather than making sense, or use *of* mathematics (or perhaps even making connections *within* the discipline), is to invoke creativity in Elliott's traditional sense, and is thus a radical claim. Phillips herself gently questions Upitis' claim about manipulatives, arguing for the potential of manipulatives as tools which not only "help the pupil to find answers (learn about mathematics)", but also "help the pupil construct mathematics, both conceptually and strategically" (Upitis *et al.*, 1997, p. 143). Such construction is, of course, 'making' – in the sense of making sense. Phillips does not develop, with examples, her claim about the use of manipulatives to support or provoke pupils' construction of mathematics, although it fairly reflects a strong undercurrent of belief within primary mathematics.

Phillips adds, significantly, that the use of manipulatives needs to be mediated. Echoing Pimm (1995, p. 13), she argues that "without mediation, the connection between the tool and the maths may not be clearly established" (Upitis *et al.*, 1997, p. 143). We would share her view that any meaning attached to the mathematics does not reside in the apparatus, but is constructed in the pupil's mind. As Fielker has written, "maybe all that children learn from manipulating blocks is an understanding of how to manipulate blocks" (1988, p. 6). If, instead, children are to learn about *mathematics* by manipulating blocks, but through the mediating influence of the teacher, then what freedom does the child have to be conceptually creative? Once again, the child's own sense-making is assured, whether or not it is deemed to be conventionally 'correct'. Anything else is uncertain, and arguably dubious. For manipulatives, or blocks, read nautilus shells, playground furniture or vehicles passing by the school gate. There *is* no mathematics in any of these, but human beings may make sense of their experiences with them in terms that we can describe as mathematical, given the sensitive mediation of a skilled (creative?) teacher who has been enabled to see these artefacts, these activities, through mathematical lenses.

Drawing these threads together, we find that not only can a pupil be creative by using his or her imagination to pose and solve problems, but he or she can to some extent be creative in the more traditional sense of making something. But it is important to underline two points here. Firstly, it is criterial that what is made is a *private* product. Making sense is a subjective notion and theories that support this way of looking at things stand precariously on the slippery slope that leads to philosophical idealism, which in its most radical form entails the thesis that the world is my idea. In this respect it breaks away from the arts that produce *public* products. Secondly, it leaves open the difficult question of whether or not making sense is tantamount to making truth.

The distinction which has just been drawn out between public and private products is an important one since it underpins one important motivation for creativity in education. In the 1950s, curricula and teaching styles emphasising compliance in student response were believed to be partly responsible for failure to compete in the world market with innovative technological advances. To ensure a steady flow of new ideas, the State looks to education to foster invention and novelty in its young. One of the most familiar examples of the demand for creativity in this sense arose from the realisation that the Russian space technology in the 1950s was overtaking that of the USA in launching the Sputnik.

If we now turn to the public *products,* those which in many case can be expected to be put to use to enable the community to flourish or civilisation to advance then we come close to what Elliott refers to as the 'new' concept of creative. Once again we must ask how far the innovation and *originality* sought by the community and State in mathematics could be left to our pupils.

Mathematics educators have come to value pupils' own methods of calculation when these bear the marks of inventiveness, even when these might not be the most efficient methods available. Pupils seldom produce mathematics that is original to the wider mathematics community, but it is often suggested that they might nevertheless produce results that are original to *them.* For example, Higginson, reflecting upon a 'paper jewels' project undertaken by some pupils, writes:

> Both ideas were original in that neither Rena nor Doug had experienced their ideas before. But neither original idea was unique. Other people make paper jewellery, and did before Rena began crafting her own. Other people have tessellated quadrilaterals as well, long before Doug discovered his pattern. Original ideas – or non-derivative ideas – can be powerful even if not unique. (Upitis *et al,* 1997, p. 91)

White (1972) has suggested that the shift from a result being "valuable within mathematics" to "being valuable for him [the pupil]" is an "obscure" notion. For one thing, the criteria for value in mathematics can be laid out [8], and for another it is not self-evident (in White's view) that for pupils to make their own 'discoveries' is more profitable than being presented with rules. In insisting that it is, we may be smuggling in our own preferred view of education. Indeed Ernest (2000) has pointed this out when he writes:

> the idea of creative personal development and the skills of mathematical questioning as a goal of schooling, ... remains trapped in an individualistic ideology that fails to acknowledge the social and societal contexts of schooling, and thus tacitly endorses the status quo. (p.11)

White himself acknowledges that one reply to his objection is that the pupil is creative because he is "thinking in the same sort of way as a creative mathematician does" (White, 1972, p. 136). White's rejoinder is that we do not know what the private processes of a mathematician are like, and even if we did this would not be significant because it is upon the products rather than the processes that we ascribe 'creative' to mathematics. The first part of this reply is not strictly true since we do at least have the testimonies of some mathematicians such as Poincaré (1956), for example, who made some claims about their thought processes. With respect to the second part of the reply, it has been pointed out by Wilson (1977, p. 112) that we simply do not say "That's a very creative symphony", "What a creative book *War and Peace* is!", and so on. He seems to have a point here. It is usually persons or perhaps minds that most naturally carry the title of 'creative' and not their public works.

White's strongest argument, however, arises from a comparison of the conditions in which the pupil and the mathematician make their respective discoveries. As he explains:

> In the case of the boy, his teacher has structured the situation in such a way, by e.g. providing clues to guide the child close to the desired goal, that the child takes it for granted that there is a rule to be discovered, that the teacher knows what it is, and that by following the teacher's direction he can come to find out what it is as well. None of these conditions hold for the creative mathematician, who therefore could not have made his discovery in the same way as the child. (p. 137)

This certainly seems to be true of what we might call 'guided discovery', but not all of the mathematical results which pupils find are initiated and drawn out in such a structured way. There might be instances of pupils' contributions which escape all of White's objections, though it must be admitted that these may not be as common as one might suppose, even from a reading of *Creative Mathematics*.

CONCLUSION

It cannot be denied that the confidence to solve practical problems throughout one's life is a commendable aim in education. Moreover, if being playful, imaginative and having a disposition to seek possibilities in or towards one's mathematics is being mathematically creative, and we have no reasons to doubt this, then creativity will almost certainly lead to more confident and successful solving of mathematical problems. And since, as we have already seen, such problem-solving is central to mathematics, then good mathematical competence will follow. In this respect mathematics could be called a 'creative discipline' (QCA, 1999).

But this does not imply that pupils are wholeheartedly making mathematics *in the same way* as mathematicians since, for one thing their scope for originality and self-expression is severely restricted in a way not characteristic of, say, the arts. Children's writing and children's paintings, for example, are not subject to the truth criterion that mathematics cannot escape without the charge of irony. Nor does it imply that the purpose of learning mathematics – the reason why we are compelled to learn it – is illuminated to any great degree by annexing creativity and mathematics. For one thing a creative approach to learning mathematics is quite compatible with a conservative opinion of the value of the mathematics learnt in this way. For another, there are no reasons to suppose that mathematics is of overriding importance as a vehicle for developing creativity outside mathematics. Much depends on the extent to which mathematics *presupposes* rather than *promotes* creativity.

Finally, if it is a substantial contribution to the question of *why* mathematics is worth teaching which is sought, then let us say that mathematics should be able to stand up for itself rather more than it would if its rationale is made to depend upon resemblance to other areas of the curriculum. Such a dependence unnecessarily shifts, rather than answers the question. In particular, to attempt to defend the learning of mathematics by annexing it too closely to the arts is misleading and ultimately fruitless. In this respect the question *Why Learn Maths?*, rather than the question of *how* the subject could or should be taught and learnt, remains unsubstantially answered by an appeal to creativity.

NOTES

1. The reference to 'discovery' is unfortunate here. As we have remarked earlier, 'discovery' is often contrasted with creativity. This is particularly true of mathematics.

2. 'In our time', BBC Radio 4, 18th January 2001

3. Some thirty of so remarks of a similar kind are contained in Moritz's (1914) anthology.

4. This is a novel evaluation of the status of the notions 'natural number' and 'prime'. The natural numbers are, arguably, a cultural inheritance from primitive times when the need to tally became evident. Few of us would recognise any part in their construction in the sense of Peano, or, indeed, in any other sense. The prime numbers, on the other hand, are self-evidently a secondary construct arising from the endowment of the natural numbers with multiplication.

5. Indeed there is something rather strained in the Craft's description of two young children working at some open sentences as 'creative arithmetic' (2000, p.77).

6. It is worth noting that Kline (1972, pp. 520-522) has deliberately tried to diminish the role of emotion in the arts in order to provide a strong re-categorisation of mathematics.

7. The presentation of Upitis, Phillips and Higginson (1997) is such that the individual voices of the three authors may be heard throughout the book.

8. Once again we can turn to Hardy for criteria of this sort. In diminishing the purely practical value of mathematics he wrote: 'The 'seriousness' of a mathematical theorem lies ... in the significance of the mathematical idea which it connects. We may say, roughly, that a mathematical idea is 'significant' if it can be connected, in a natural and illuminating way with a large complex of other mathematical ideas ... a serious mathematical theorem ... is likely to lead to important advances in mathematics itself and even in other sciences' (Hardy, 1967,. p. 89)

REFERENCES

Aristotle: 1955 edn, *Ethics*. Harmondsworth: Penguin.

Bramall, S.: 2000, 'Rethinking the place of mathematical knowledge in the curriculum'. In Bramall, S. and White, J. (eds.):*Why Learn Maths?* (pp. 48-62) London: University of London Institute of Education.

Bramall, S. and White, J. (eds.): 2000, *Why Learn Maths?* London: University of London Institute of Education.

Burton, L.: 1999, 'The practices of mathematicians: what do they tell us about coming to know mathematics', *Educational Studies in Mathematics* 37(2), 121-143

Craft, A.:2000, *Creativity Across The Primary Curriculum*, London: Routledge.

Dawson, S.:1991, 'Learning mathematics does not (necessarily) mean constructing the right knowledge'. In Pimm, D. and Love, E. (eds), *Teaching and Learning School Mathematics*, London Hodder and Stoughton, pp. 195-204.

Dearden, R. F.: 1967, *The Philosophy of Primary Education.* London: Routledge and Kegan Paul.

DES: 1982, *Mathematics Counts.* London: HMSO.

Elliott, R. K.: 1971, 'Versions of creativity', *Proceedings of the Philosophy of Education Society of Great Britain,* V(2), 139-151.

Ernest, P.: 2000, 'Why teach mathematics?'. In Bramall, S. and White, J. (eds.): *Why Learn Maths?* (pp. 1-14), London: University of London Institute of Education.

Fielker, D.: 1988, 'Metaphors and models'. *Mathematics Teaching* 124, 4-6.

Gregory, R. L.:1987 (ed), *The Oxford Companion to the Mind.* Oxford: Oxford University Press.

Hardy, G. H: 1967, *A Mathematician's Apology.* Cambridge: Cambridge University Press.

Huckstep, P.:2000, 'Mathematics as a vehicle for 'mental training''. In Bramall, S. and White, J. (eds.) *Why Learn Maths?* (pp. 85-100), London: University of London Institute of Education.

Kline, M: 1972, *Mathematics in Western Culture.* Harmondsworth: Penguin.

Moritz, R. E.: 1914, *On Mathematics and Mathematicians.* New York: Dover.

Pimm, D.: 1995, *Symbols and Meanings in School Mathematics.* London: Routledge.

Popper, K. R.: 1972, *Objective Knowledge.* Oxford: Oxford University Press.

Poincaré, H: 1956, 'Mathematical Creation'. In J. R. Newman (ed.) *The World of Mathematics* (pp. 2041-2050). New York: Simon and Schuster.

Prowse, M.: 2000, 'Dropping maths? It just doesn't add up'. *Financial Times,* 14th October 2000

QCA: 1999, *The National Curriculum.* London: DfEE.

Rowland, T.:1999, *The Pragmatics of Mathematics Education: vagueness and mathematical discourse.* London: Falmer.

Sawyer, W. W.: 1955, *Prelude to Mathematics* Harmondsworth: Penguin.

Sullivan J. W. N.: 1960, 'Mathematics as an art' In Newman, J. R. (ed) *The World of Mathematics* Vol. 3 (pp. 2015-2021) London: George Allen and Unwin.

Upitis, R., Phillips, E. and Higginson, W.: 1997, *Creative Mathematics: Exploring Children's Understanding.* London: Routledge.

White, J. P.: 1972, 'Creativity and education: a philosophical analysis'. In R. F. Dearden, P. H. Hirst and R. S. Peters (ed.) *A Critique of Current Educational Aims* (pp. 130-146). London: Routledge and Kegan Paul.

Wilson, J.: 1977, *Philosophy and Practical Education.* London: Routledge and Kegan Paul.

THEME 4

LEARNING MATHEMATICS WITH TECHNOLOGY

It is, perhaps, inevitable that a collection of current research in mathematics education should include a section related to the use of technology. The chapters included here, however, do rather more than simply evaluate the effects of introducing technology. Each looks deeply at the ways in which students learn in specific technology-rich environments and at the nature of the mathematical thinking that occurs.

Geometry and proof are both contested areas of the UK school mathematics curriculum and are currently seen to be in need of development. Federica Olivera contributes to this development by exploring ways in which using the dynamic geometry software Cabri-Géomètre can provide support for conjecturing and proving in geometry. Her chapter describes a teaching experiment in which pairs of students worked on open problems using Cabri. Analysis of their problem solving processes focuses on the ways the students used the possibilities the software provides for 'dragging' geometric objects. The role this played in their production of conjectures is explored.

Making links between visual and algebraic ways of thinking is one of the important goals of mathematical learning that new technology, with its potential for manipulation of both symbols and graphics, may be able to support. Soo Duck Chae and David Tall investigate this aspect of the learning of undergraduate mathematics students during a course in which they experimented in a graphical software environment. The evidence they present suggests that many of the students succeeded in making connections between numerical, symbolic and visual representations as they gained insight into the conceptual foundations of chaos theory.

As well as the cognitive aspects of learning with technology, it is important to consider the social context within which learning takes place and the role that technology may play in this. Socio-cultural theories of learning prioritise the interpersonal nature of learning and are sometimes used as an argument in support of group work in the classroom. Sally Elliott, Brian Hudson and Declan O'Reilly use these ideas to analyse the work of a small group of students using graphical calculators. They examine the ways in which shared knowledge is constructed during a task of matching graphs to algebraically expressed functions, the role the calculators play in this and interactions between members of the group, concluding that it is useful to conceive of the group as a "local community of practice".

12 CONSTRUCTION OF CONCEPTUAL KNOWLEDGE: THE CASE OF COMPUTER-AIDED EXPLORATION OF PERIOD DOUBLING

Soo D. Chae and David O. Tall

Mathematics Education Research Centre, University of Warwick

This research focuses on students using an experimental approach with computer software to give visual meaning to symbolic ideas and to provide a basis for further generalisation. They use computer software that draws orbits of x = f(x) iteration and are encouraged to investigate the iterations of $f_\lambda(x)=\lambda x(1-x)$ as λ increases. The iterations pass through successive acts of period-doubling as $\lambda=\lambda_0,\ \lambda_1,\ \lambda_2,\ ...;$ students are invited to estimate the values of λ and to compare their experimental results with the theory of geometric convergence. The supervisor acts as a mentor, using various styles of questioning to provoke links between different ideas. A variety of data is collected to give evidence for the ways in which students develop conceptual links between symbolic theory and the visual and numeric aspects of computer experiment.

INTRODUCTION

According to Dreyfus and Eisenberg (1991), many students prefer to think algebraically rather than geometrically when they are solving problems, and the authors give several reasons for this in terms of social, curricular and epistemological factors. For example, students are more likely to solve the equation $f(x) = x$ rather than drawing a graph of the function and a diagonal line when asked to find fixed points of the given function f. In other words, they think algebraically about a given problem using equations, symbols and logic rather than draw a diagram or pictures. On the other hand, some people are visualisers, who prefer visual and spatial methods and think geometrically when they are asked to solve a problem. Research concerned with the use of visual abilities reveals not only the value of visual processing in visuo-spatial problems, but also that both high visualisers and low visualisers can improve their understanding with computer-assisted learning using graphical representations (Sein, Olfman, Bostrom and Davis, 1993).

> Restoring the visual and intuitive side of mathematics opens new possibilities for mathematical work, especially now that computing has enough power and resolution to support it with accurate representations of problems and their solutions. The benefits of visualization include the ability to focus on specific components and details of very complex problems, to show the dynamics of systems and processes, and to increase intuition and understanding of mathematical problems and processes (Cunningham, 1991, p. 70).

Tall (1991a) also observes that computer graphic software can provide students with environments for intuition prior to the construction of a formal concept. With these ideas in mind, we followed a course in which a mathematics professor provided students with a software environment to "get their hands dirty" through an experimental approach that encouraged them to think visually and numerically rather than just symbolically. These students were first year mathematics majors in a university with a high quality student intake (all with A-grades at A-level mathematics and a minimum of AAB in three subjects). In their first two terms they had received lectures and seminars on mathematical analysis, differential equations, linear algebra, and group theory. They were now attending a computer laboratory course in which they were to extend their symbolic experience with visual explorations to lay conceptual foundations for chaos theory.

Sierpinska (1987) has warned that the students may perceive the pictures arising in such a software environment in an immediate, intuitive and global way, subtle ideas being obscured in the potential infinity of the symbolic process. If the picture on the screen converges quickly, it may seem that it is actually attained, while, if it converges slowly, the student may not believe that it converges at all. We investigate whether this occurs and see what kind of arguments the students use to link experimental numerical results and formal theory. Do the visuo-spatial experiences with the computer, supported by the supervisor as mentor, provide a basis for reflective activities that lead to flexible conceptual thinking relating numerical experiments, visual representations and symbolic theory?

> Experience modifies human beliefs. We learn from experience or, rather, we ought to learn from experience. To make the best possible use of experience is one of the great human tasks and to work for this task is the proper vocation of scientists. (Polya, 1954, p. 3).

PROCEDURAL AND CONCEPTUAL KNOWLEDGE

Hiebert and Lefevre (1986, p. 3) state that the crucial characteristic of conceptual knowledge lies in the rich relationships constructed between specific pieces of information. It may be considered as a well-connected web of knowledge for flexibly accessing and selecting information. In contrast, procedural knowledge is a form of sequential knowledge, constructed in a succession of steps.

Heid (1988) showed that students in an experimental calculus class using a microcomputer as a tool for visualizing graphs and for manipulating symbols developed a broader conceptual understanding than students in a traditional class focusing mainly on symbolic procedures. She found that students gaining conceptual knowledge in this way were able to develop concepts further than those using procedural knowledge. Many other researchers (e.g., Tall, 1991b) contend that students using interactive dynamic software gain a much better insight into mathematical concepts than those following a traditional curriculum. In this study we therefore consider conceptual knowledge constructed through visualisation using interactive graphical software. The present research was conducted using the framework outlined in the following sections, focusing on the establishment of the connection between visual orbits of $x = f(x)$ iteration, the numeric information provided by the software and the underlying mathematical theory.

THE MATHEMATICAL CONTEXT

The research focuses on a mathematical activity which is part of a theoretical and experimental development leading to chaos theory. The computer class was preceded by a one hour guided symbolic investigation in which the students investigated the fixed points of $f(x) = \lambda x(1-x)$ symbolically. This involved solving the equation $x = \lambda x(1-x)$ for x in terms of λ to obtain the roots in symbolic form, namely 0 and $1-1/\lambda$. Students then investigated the size of $f'(x)$ at these fixed points. If $|f'(\alpha)| < 1$ at a fixed point α, then α is an *attractor*, and iterations will home in on it. On the other hand, if $|f'(\alpha)| > 1$ then α is a *repeller* and iterations will move away. By symbolic means the student is expected to determine when the fixed point $1-1/\lambda$ is an attractor. Since $f'(x) = \lambda - 2\lambda x$, it is easy to show that $x = 1-1/\lambda$ is an attractor for $1 < \lambda < 3$.

The student is then invited to carry out similar calculations for the function $f^2(x) = f(f(x))$ which is a little more intricate but possible. (The students involved are very able and 14 out of 19 were able to complete the symbolic task.) The calculations for higher iterates $f^{n+1}(x) = f(f^n(x))$, however, become more complex and at this point it is time to switch to the computer model.

The student is invited to investigate iteration of the function $f(x) = \lambda x(1-x)$ as the parameter λ increases, using computer software. For values of λ between 1 and 3, the iterations home in on the attractive root (figure 1(a)). Although this contains the seeds of the Sierpinska obstacle (that the limit may *actually* be reached visually, but not symbolically), the visual picture allows the encapsulation of the limiting process as a visual limit object, the *point x* where $x = f(x)$ iteration stabilizes. This point may then be seen to vary, changing smoothly as λ increases. When λ passes through the value 3, the attractive point becomes a repeller and the iterations begin to spiral out and settle in a period of length two (figure 1 (b)). This phenomenon is called a *period doubling bifurcation*.

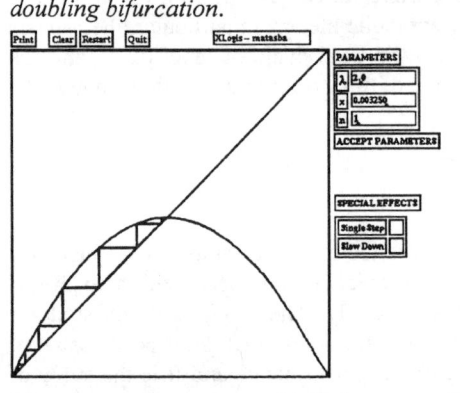

(a) convergence to one root of $f(x) = x$ (b) period doubling after $\lambda=3$ (here $\lambda=3\cdot2$)

Figure 1: Graphic representations using the software *xlogis*

As λ continues to increase, at $\lambda = \lambda_1$, a new bifurcation occurs for f, from a cycle of period 2 to one of period 4. This corresponds directly to a simpler bifurcation for f^2 from a fixed point to a cycle of period 2. (Figure 2.)

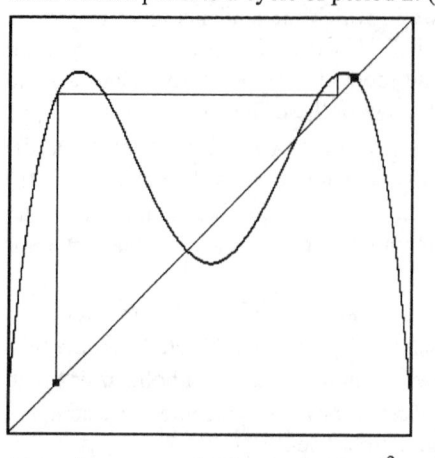 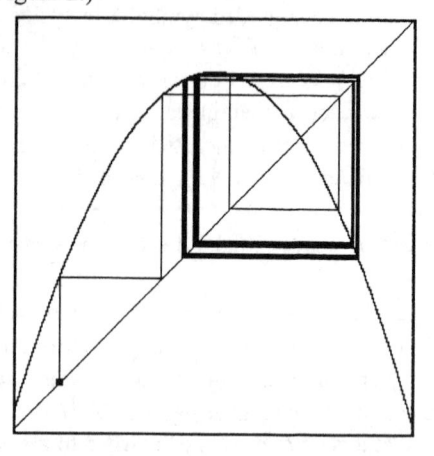

(c) as λ increases, the fixed point for f^2 becomes unstable and bifurcates

(d) The corresponding cycle of period 2 for f opens out to a cycle of period 4

Figure 2: The second bifurcation

As λ increases further, there are a successive bifurcations of period 2, 4, 8, and so on, at values of $\lambda = \lambda_0$, λ_1, λ_2. The purpose of the computer investigation is to get numerical data on the first few values of this sequence and to get a sense that they satisfy the condition for geometric convergence. The sequence, λ_0, λ_1, λ_2 ... converges to a value λ_∞ called the Feigenbaum point. The computer experience is therefore intended to give the student an experimental context offering visual pictures and numerical approximation to link to the symbolic theory. In particular, the *process* of iteration can be seen as a visual *object* – the final cycle of length 2^n – and the student can imagine the successive behaviours of this object as λ increases to give the succession of bifurcations.

THE RESEARCH FRAMEWORK

Subjects

The study involved thirty first-year students enrolled in an Experimental Mathematics course at the University of Warwick. The first-named author supervised three groups containing five, six, and seven students, respectively. The group with six students consisted entirely of students who had obtained first-class marks on previous tests. The other two groups had a broader spectrum of performance and it is the study of these two groups that is reported in this paper.

Software: *xlogis*

The software used was *xlogis* (see figure 1(a)). This software is written in C and runs under the Xwindows Graphic User Interface at Sun terminals operating Sun Solaris. It is designed to enable students to control a range of parameters for iterating the function $f(x) = \lambda x(1 - x)$; these include options to specify the value of λ, the starting point x and the number n for the display of the nth iterate $f^n(x)$. In addition there are options that allow the user to operate the iteration one step at a time or to change the pace of the iteration. The iterations drawn so far may also be cleared, enabling the user to focus on the later iterations when they have stabilized on a limit cycle.

Instruments

Various forms of data were collected in the study. The course organizer had already designed a *pre-requisite test* to test the students' understanding of "geometric convergence" (see below). The formal assessment for the course consisted of a *written assignment* handed in three days after each session, requiring students to write about their observations and inferences. In addition to these formal assessments, Soo D. Chae, who acted as supervisor and participant-observer, collected data using *audio-tapes* together with *field notes* made at the time. Finally, the students were given a *questionnaire* after the course to investigate some of their understandings relating to their visual experiences and symbolic theory.

Pre-requisite test

This was designed to investigate students' awareness of "geometric convergence" given in terms of the following definition and the accompanying question:

A sequence (a_n) is said to *converge geometrically* if the ratio $(a_{n+2} - a_{n+1})/(a_{n+1} - a_n)$ converges to a limit r with $0 \leq r < 1$ as n goes to infinity. Write down an example of a sequence that converges geometrically. Prove that a sequence which converges geometrically also converges in the usual sense.

The computer experiment

The students were given the following tasks to experiment with the logistic map $f(x) = \lambda x(1 - x)$.

1. Use *xlogis* to investigate what happens when λ increases through the value 3·0.

2. Use *xlogis* to investigate the dynamics for λ between 3 and the value λ_1 for which the period 4 orbit occurs. What happens when λ goes through λ_1?

3. As you increase λ beyond λ_1, you should see a sequence of period doubling bifurcations. Use *xlogis* to obtain estimates of the parameter values λ_n for which the nth period doubling bifurcation occurs.

What do you notice about the way the λ_n converge? The parameter value λ_∞ to which they converge is called the accumulation of period doublings. Try taking the ratios of successive differences. What does the result tell you? Can you think of a way of seeing this by drawing a graph?

203

In this sequence of activities, the students begin with a value of λ less than 3 to reveal the picture in figure 1(a), but as the value of λ passes through $\lambda_0 = 3$, the picture changes to the format of figure 1(b). They must then experiment with larger values of λ to estimate the values $\lambda = \lambda_1$, $\lambda = \lambda_2$, ..., where successive period doublings occur. These must be performed as accurately as possible to be able to relate them to the theory of geometric convergence. In practice the accuracy is limited as it involves the student trying various values of λ and homing in on the points where the orbits change; the exact point where the change occurs can only be seen approximately on the computer screen.

The role of the supervisor

In order to improve effective experimentation, the supervisor assisted and responded to the group, providing support and explaining the phenomenon of period doubling. Sometimes the supervisor offered advice by providing directed questions to keep the students going if they were stuck. Three different types of questions were used: for *opening-up*, *structuring*, and *checking* (Ainley, 1988). For instance, an *opening-up question* responds to a student's request by asking the student to think more about it:

Student A: What is happening when the function cycles between two values? (referring to the picture in figure 1(b)).

Supervisor: How does this relate to the terms of the sequence x_1, $x_2 = f(x_1)$, $x_3 = f(x_2)$, ...?

A *structuring question* is designed to construct concepts by linking disconnected knowledge via appropriate structured directing questions:

Student B: This equation is quite complicated to solve. How can I find the solutions? (pointing to the fourth degree equation generated by $f^2(x) = x$).

Supervisor: Do you think that fixed points of $f(x)$ also become fixed points of $f^2(x)$?

Student B: Maybe.... Umm .... Yes....

Supervisor: Why? Justify your answer. (Prompting the student to make links explicit.)

A *checking question* simply checks what the student has just done:

Student C: The function seems to be hitting four points. So, is this lambda one?

Supervisor: Will a slightly smaller value of lambda also hit four points?

Students' self-written reports

Students were asked to write up their observations and answers as they proceeded, and then to summarise their mathematical ideas and arguments clearly and hand them in within three days. According to Mason (1982), this kind of activity is valuable for helping students to reflect on what they have done and how they have done it. The supervisor graded the reports using criteria that emphasised the quality of students' ideas without seeking perfect presentation. The students' reports on the mathematical questions posed during the experimentation provided a valuable source of data.

RESULTS AND DISCUSSION

Student responses to the pre-requisite test

Of the twelve students in the two groups studied, two did not give responses to the request for an example of geometric convergence, three gave incorrect responses (a_n =1/n, $\sqrt{n}$) and seven gave correct examples (a_n=2^{-n}, or 10^{-n}, or e^{-n}). Only five students could give a proof that geometric convergence implies convergence in the usual sense. (See the first two columns of Table 2 below.) Thus less than half of the students responded successfully to what was considered a necessary pre-requisite for this course.

Students concept images apparent in the post-course questionnaire

The post-course questionnaire included a question (figure 3) which gave the definition of a fixed point and asked the student to identify the fixed points of the iteration $x = f(x)$ in a picture. Despite the definition stating that it is a point x such that $y = f(x)$, the visual representation focuses instead on the point (x, y) where the line $y = x$ meets the curve $y = f(x)$.

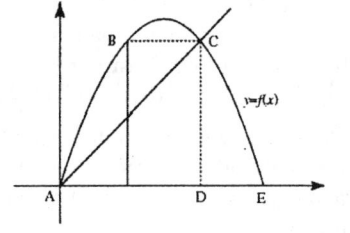

Figure 3: Which points are fixed points of the given function?

As we see from Table 1, only one student answered correctly while most students chose the intersection points A and C on the diagonal line. This is one example of a geometric obstacle. In this case it proved to be easy to reconcile the definition with the picture through discussion.

Case	Number (%)	Concept Image
A & C	18 (60%)	A fixed point of f is where the graph of f intersects the diagonal
A & D	1 (3%)	*[the correct response]*
A, C, & E	1 (3%)	
A, B, C & E	1 (3%)	D is not on the graph $y=f(x)$
C	2 (6%)	
E	1 (3%)	
No response	6 (20%)	

Table 1: Concept images for fixed points

The graphical obstacle

The major obstacle referred to by Sierpinska—in which the screen picture is interpreted differently from the mathematical theory—is more subtle. When the students investigated the convergence for $\lambda = 3$, almost all the students concluded that it converged slowly; only one concluded it formed a tiny cycle of length 2. (In a later experiment, more students believed that the iteration failed to converge.)

In speaking of the convergence of $\lambda_0, \lambda_1, \lambda_2, \lambda_3, \ldots$, student A was earlier quoted asking 'What happens when the function cycles between two values'; student C said 'the function *seems* to be hitting four points.' This language allows the limit cycles of period 1, 2, 4, ... to be viewed as *mental objects* that change as λ increases. This actually *helps* the students to view the period doubling phenomenon occurring again and again at $\lambda_0, \lambda_1, \lambda_2, \lambda_3, \ldots$. This final convergence to the Feigenbaum point was related to geometric convergence by nine of the twelve students. In their written reports and interviews more detail about their conceptions emerged.

Students' formulations of period doubling

The students' written reports were analysed to see how they responded to the tasks given them. One student, typical of those who were successful, estimated the numerical sequence of period doubling bifurcations as approximately $\lambda_0=3$, $\lambda_1=3\cdot449$, $\lambda_2=3\cdot54$, $\lambda_3=3\cdot559$, $\lambda_4=3\cdot563$, $\lambda_5=3\cdot564$, $\lambda_6=3\cdot5642$. He observed:

> The values for λ_n appear to be converging to a value between 3·5 and 3·6. The parameter value λ_∞ to which the λ_n's would converge is called the accumulation of period doublings. Taking the ratio $(\lambda_n - \lambda_{n-1})/(\lambda_{n-1} - \lambda_{n-2})$ of successive differences, we find 0·2027, 0·2088, 0·2105, 0·25, 0·21.

He noted that—despite the poor accuracy afforded by the limitations of the experiment—the values of the ratio were initially around 0·2, which is far less than the critical value 1 for geometric convergence.

Our main concern is the manner in which each student coped with the convergence of the sequence in the written assignment (table 2). Student S1 attempted to give a symbolic proof of the convergence, in addition to drawing a graph of his numeric computations, plotting values of $\lambda_n - \lambda_{n-1}$ against $\lambda_{n-1} - \lambda_{n-2}$ to reveal a line of gradient approximately 0·2. Student S2 drew a similar graph together with a rough symbolic argument. Students S3 and S4 plotted the values of $(\lambda_n - \lambda_{n-1})/(\lambda_{n-1} - \lambda_{n-2})$ for λ against n to obtain a sequence of points approximating to the horizontal line $\lambda=0\cdot2$. Student S5 obtained the numerical values and simply observed that they were approximately 0·2. Students S6 and S7 were not able to obtain satisfactory numerical values to be able to attempt the task.

Nine students (S1 to S4, S8 to S12) were able to give some kind of explanation relating the experimental results to the theory of geometric convergence. This included S3, S4 and S8, who had not responded satisfactorily to the pre-requisite test.

This simply means that these three students were able to *use* the test for geometric convergence without being able to give examples or to give a formal proof that geometric convergence implies convergence.

	Pre-requisite Test		Written Assignment			
	Geometric convergence	Proof	Symbolic	Numeric	Graphic	Geometric convergence to λ_∞
S1	yes	yes	yes	yes	yes	yes
S2	yes	no	yes	yes	yes	yes
S3	no	no	yes	yes	yes	yes
S4	no	no	yes	yes	yes	yes
S5	no	no	yes	yes	no	no
S6	yes	no	no	no	no	no
S7	no	no	no	no	no	no
S8	no	yes	yes	yes	yes	yes
S9	yes	yes	yes	yes	yes	yes
S10	yes	yes	yes	yes	yes	yes
S11	yes	no	yes	yes	yes	yes
S12	yes	yes	no	yes	no	yes

Table 2: Responses of observed groups in the pre-requisite test and written assignment

Nine of the twelve students (S1 to S5, S8 to S11) succeeded in relating their experiences to their earlier use of symbolism in the preliminary work. Interestingly, all of these students also succeeded in using numeric representations in the written part of the assignment. More importantly, eight students (S1 to S4, S8 to S11) out of these nine were able to also provide a graphic representation of their numerical data and to link this to the geometric convergence of the sequence of λs to the Feigenbaum point. These findings underline the fact that two thirds of the students involved were successful in connecting their visual and numeric observations to theoretical aspects of the situation, intimating that the experiment was successful in aiding their construction of wider conceptual knowledge.

SUMMARY

This investigation into students using computer software to gain visual insight and to obtain numerical approximations to link with theory proved to be successful for eight out of twelve students, including three who did not have the desired pre-requisite knowledge of the notion of geometric convergence. These three were able to operate by simply substituting into the given formula $(\lambda_n - \lambda_{n-1})/(\lambda_{n-1} - \lambda_{n-2})$. One minor cognitive obstacle encountered by the majority of students is that the picture gives the impression that the fixed points are where the curve $y = f(x)$ meets the line $y = x$ rather than the value of x for which $f(x) = x$. This was easily resolved.

Several students used language that intimated that the limiting cycles of period 2, 4, etc., represented the end result of the process of iteration, allowing them to speak of the cycles as mental objects yielding a sequence of bifurcations as λ increased. This seems to be a perfectly natural process of *encapsulating the process of iteration* to give *visual objects that could be mentally manipulated*, without necessarily falling into the Sierpinska obstacle of equating what was on screen precisely with the underlying, potentially infinite, mathematical processes.

Overall, two thirds of these students were able to use flexible links between numeric, graphic and symbolic representations of geometric convergence to construct their own ideas of the convergence of the points of bifurcation to the Feigenbaum point. Thus the claim by Dreyfus and Eisenberg that students prefer to think algebraically rather than geometrically should not be interpreted to mean that students *never* think geometrically. By giving students environments in which flexible thinking is encouraged, flexible thinking relating numeric, visual and symbolic representations can – and does – occur.

REFERENCES

Ainley, J.: 1988, 'Perceptions of teachers' questioning styles.' In A. Borbas (ed.), *Proceedings of the 12th Conference of the International Group for the Psychology of Mathematics Education* (Vol. 1, pp. 92-99). Vesprem, Hungary.

Cunningham, S.: 1991, 'The visualization environment for mathematics education.' In W. Zimmerman and S. Cunningham (eds.), *Visualization in Teaching and Learning Mathematics* (pp. 67-76). Providence, RI: MAA Notes No. 19.

Dreyfus, T. and Eisenberg, T.: 1991, 'On the reluctance to visualize in mathematics.' In W. Zimmermann and S. Cunningham (eds.), *Visualization in Teaching and Learning Mathematics* (pp. 25-37). Providence, RI: MAA Notes No. 19.

Heid, M. K.: 1988, 'Resequencing skills and concepts in applied calculus using the computer as a tool.' *Journal for Research in Mathematics Education, 19*(1), 3-25.

Hiebert, J. and Lefevre, P.: 1986, 'Conceptual and procedural knowledge in mathematics : an introductory analysis.' In J. Hiebert (ed.), *Conceptual and Procedural Knowledge: The Case for Mathematics* (pp. 1-27), Hillsdale, NJ: Lawrence Erlbaum Associates.

Mason, J., with Burton, L. and Stacey, K.: 1982, *Thinking Mathematically*. London: Addison-Wesley.

Polya, G.: 1954, *Induction and Analogy in Mathematics*. Princeton, NJ: Princeton University Press.

Sein, M. K., Olfman, L., Bostrom, R. P., Davis, S. A.: 1993, 'Visualisation ability as a predictor of user learning success.' *International Journal of Man-Machine Studies*, *39*, 599-620.

Sierpinska, A.: 1987, 'Attractive fixed points and humanities students.' In J. Bergeron, N. Herscovics, and C. Kieran (eds.), *Proceedings of the 11th International Conference for the Psychology of Mathematics Education* (Vol. 3, pp. 170-176). Montreal.

Tall, D. O.: 1991a, 'Intuition and rigour: the role of visualization in the calculus.' In W. Zimmermann and S. Cunningham (eds.), *Visualization in Teaching and Learning Mathematics* (pp. 25-37). Providence, RI: MAA Notes No. 19.

Tall, D. O.: 1991b, 'Recent developments in the use of the computer to visualise and symbolise calculus concepts.' In C. Leinbach (ed.), *The Laboratory Approach to Teaching Calculus* (pp. 15-20). Providence, RI: MAA Notes No. 20

13 HOW DOES THE WAY IN WHICH INDIVIDUAL STUDENTS BEHAVE AFFECT THE SHARED CONSTRUCTION OF MEANING?

Sally Elliott, Brian Hudson, Sheffield Hallam University

Declan O'Reilly, University of Sheffield

Audio taped discussions between three students have been examined to shed light on the way in which the behaviour of individual students may affect the shared construction of meaning with graphical calculators. These discussions revealed a complex pattern of interaction between the students. Each student was responsible for defining his or her own role within the discourse and these roles appeared to change as the discussion progressed. With reference to the framework offered by Winbourne and Watson (1998), it is proposed that local communities of practice have been established and that the individual student's positioning within the community of practice determines their success as a learner and contributes towards the creation of shared knowledge.

BACKGROUND

The study reported in this paper seeks to investigate whether three GCE (General Certificate of Education) Advanced level Further Mathematics students were able to develop a joint conception of the problems that they worked on together. They discussed as a group, with graphical calculators available. Of particular interest was the part that each individual student played in creating shared meaning in the context of the technology. The theoretical position adopted in this study is based on the Vygotskian idea that all learning is essentially social and is mediated by tools. Meaning is derived through interactions between students and with the teacher. Each participant occupies a different role in the construction and negotiation of meaning and these roles are developed through participation in local communities of practice. These ideas, which form the basis for this study, are elaborated below and discussions between students and teacher working on a mathematical task are then analysed using these theoretical constructs.

Socio-Cultural Learning

Vygotsky proposed that all individual mental processes are based on social interactions. Interactions experienced within the social context are internalised by the individual and learning proceeds from the interpsychological to the intrapsychological. Furthermore, the learning process is mediated by the use of tools, such as speech, symbols, writing and technology. Within a Vygotskian perspective, tools are seen to fundamentally shape and define activity. They are used firstly as a means of communicating with others, to "mediate contact with our social worlds", and eventually "these artifacts come to mediate our interactions with self; to help us think, we internalise their use" (Moll, 1990, p. 11-12). In particular, Vygotsky regarded language as the means through which thought is developed: "thought is not merely expressed in words; it comes to exist through them" (*ibid.*, p. 125).

The site in which learning takes place is the *zone of proximal development (ZPD)*. Vygotsky defined this as "the distance between the actual developmental level as determined by independent problem solving and the level of potential development as determined through problem solving under adult guidance or in collaboration with more capable peers" (Vygotsky, 1978, p. 86). Individuals learn from interaction with other more knowledgeable persons in the *ZPD*. Consequently peer tutoring, peer collaboration and teacher intervention play an important part in constituting the *ZPD*.

Social Construction of Meaning

In developing a Vygotskian perspective, Lerman (1994) regards meaning as socio-cultural in nature – a product of discourse and discourse positions. Individuals are acculturated into meanings and thus the intersubjective becomes the intrasubjective. The individual student's input into meaning-making changes and is changed by the discourse. In this way the student derives meaning from their *positioning* in social practices. Meaning is *appropriated* by individual students, whereby each student forms his or her own something, from that which already belongs to other people. This appropriation occurs through communication and tool use.

As concepts derive their meaning from being used, the acquisition of a concept or understanding can be interpreted as the result of an individual coming to share in that meaning through negotiation and discussion (Lerman 1996). Mathematical concepts are social acts and tools; as these concepts are socially determined, they are socially acquired. Jones & Mercer (1993) propose that successful learning occurs when two or more people manage to share their knowledge and understanding, so that a new cultural resource is created which is greater than the knowledge and understanding any of the individuals hitherto possessed. They stress that much learning, not least in relation to information technology, consists of sharing knowledge.

Local Communities of Practice

The study reported here was concerned with creating a classroom environment that would facilitate and support the negotiation of meaning between students and teacher, thereby giving rise to successful learning opportunities. The teacher-researcher deliberately set out to establish *local communities of practice* in order to achieve this aim. Winbourne and Watson (1998) identify six key features necessary for initiating *local communities of practice (LCP)*:

1. Pupils see themselves as functioning mathematically within the lesson;

2. There is a public recognition of competence;

3. Learners see themselves as working together towards the achievement of a common understanding;

4. There are shared ways of behaving, language, habits, values and tool-use;

5. The shape of the lesson is dependent upon the active participation of the students;

6. Learners and teachers see themselves as engaged in the same activity.

They propose that any classroom can be regarded as an intersection of a multiplicity of practices and trajectories. They further argue that the individual student's *positioning* within the community of practice will determine their success as a learner. Ultimately, the students can come to operate masterfully, within the constraints of the social setting. The process by which the individual achieves his or her position within a community of practice is explained by the notion of *telos*. This notion presupposes a common direction of learning and Winbourne and Watson broadly describe telos as "an unfulfilled potential to move or change in many different ways" (p. 182). They contend that "telos could be conceptualised as a set of constraints in some sense inherent in situations and in the individual's pre-dispositions to respond to situations as she does" (p. 182). In this sense the individual student's learning is both determinant of the common direction of learning and in part determined by the complex paths that the students have taken to be where they are. The students fulfil their ultimate positions within the community of practice through smaller-scale 'becomings' in which they join the practice and begin to assume their eventual position. For example participation in the practice of asking questions can enable students to generate mathematical questions themselves. Similarly, participation in the practice of using graphical calculators can allow students to become "masters" in the use of these tools. The students' experiences at school are mediated by the images of themselves as learners that they bring with them.

The Role of Technology

In exploring how the use of technology mediates students' learning of mathematics, Pea distinguished between the *amplification* and the *cognitive reorganisation* effects of technology (cited in Berger, 1998). The *amplification* effects refer to the speed and ease by which the student is able to operate whilst using the technology. In this study, the graphical calculator is seen to *amplify* the zone of proximal development by creating a situation where the student is able to complete more conceptually demanding tasks effectively and easily. The benefits of *amplification* are regarded as short-term phenomena, providing the student with immediate assistance during problem solving.

Use of the graphical calculator may also enrich or change students' conceptions in some way and thus may function as a tool which helps the students' thinking to develop. This is referred to as the *cognitive reorganisation* effect of the technology, which is defined by Berger as "a systematic change in the consciousness of the learner, occurring as a result of interaction with a new and alternate semiotic system" (1998, p. 16). Long-term changes in the quality of learning arise through *cognitive re-organisation*. Berger argues that the learner needs to engage thoughtfully with the technology if internalisation is to occur. It is not sufficient for a student merely to be introduced to the technology. Berger further suggests that in order for the learner to "interact in such a mindful way" he or she needs to "use the technology actively and consciously in a socially or educationally significant way" (1998, p. 19).

In a study conducted by Borba (1996), the use of the graphical calculator was seen to "enhance mathematical discussions" and this in turn "reorganised" the way in which knowledge was constructed in the classroom. In such a social context, the graphical calculator is seen as a mediator of both the teacher-student relationships and the interactions between individual students. Pea (1987) argues that "social environments that establish an interactive social context for discussing, reflecting upon, and collaborating in the mathematical thinking necessary to solve a problem also motivate mathematical thinking" (p.104). Furthermore, he also emphasises that technology can play a "fundamental mediational" role in promoting dialogue and collaboration in mathematical problem solving.

The Role of the Teacher

In a Vygotskian framework both the teacher and the students play a mutual and active part in creating the social environment (Moll, 1990). The function of the teacher is seen as an integral part of any learning situation. To discuss teaching and learning separately would thus make no sense from a Vygotskian viewpoint. The teacher is seen as a mediator of student learning and assumes an active and necessary role in the learning process (Lerman, 1994). An important objective for the teacher is to apprentice students into the discourse of the mathematics classroom. The teacher assists the students in "appropriating the culture of the community of mathematicians as a further social practice" (Lerman, 1996, p. 146). Consequently the students will be able to operate masterfully in this setting. Likewise Moll (1990) argues that a major role for the teacher is in creating social contexts for mastery of and conscious awareness in the use of cultural tools. By "constraining the foci for attention, and by recognising and working with pre-dispositions, rather than ignoring them, a teacher is more likely to be able to initiate *local communities of practice* which enable learners to see themselves as members of a mathematical community" (Winbourne and Watson, 1998, p. 183).

The theoretical framework discussed above illustrates that interaction between peers, with the teacher, and with technology, within a supportive learning environment are key elements in students' meaning-making in the mathematics classroom. It also raises some important questions – particularly concerning the nature of shared knowledge. For example, when can knowledge be taken as shared? What role does each individual play in constructing such knowledge and how is this then "appropriated" (Lerman, 1994)? How does the teacher or use of the technology facilitate the appropriation process? How does the individual student make further use of shared knowledge? This study has sought to address these issues and in doing so has attempted to elaborate further on the complex process of knowledge acquisition in the mathematics classroom.

METHODOLOGY AND DATA COLLECTION

The work reported in this paper forms part of a broader study of the way in which the graphical calculator mediates students' learning of functions, and the data was collected during the second phase of this research (see Elliott, 1998 for details on the first phase). The methodological approach adopted in this study is both qualitative and ethnographic and is based on the underlying assumption that "all human activity is fundamentally a social and meaning making experience" (Eisenhart, 1988, p. 102). The principles of ethnography have thus governed the whole approach to carrying out the classroom-based research to date, from the choice of methods of enquiry, to the way in which each episode has been interpreted within the world of the participants. For a more detailed overview of this phase of the study and the way in which data was collected see Elliott, Hudson and O'Reilly (2000).

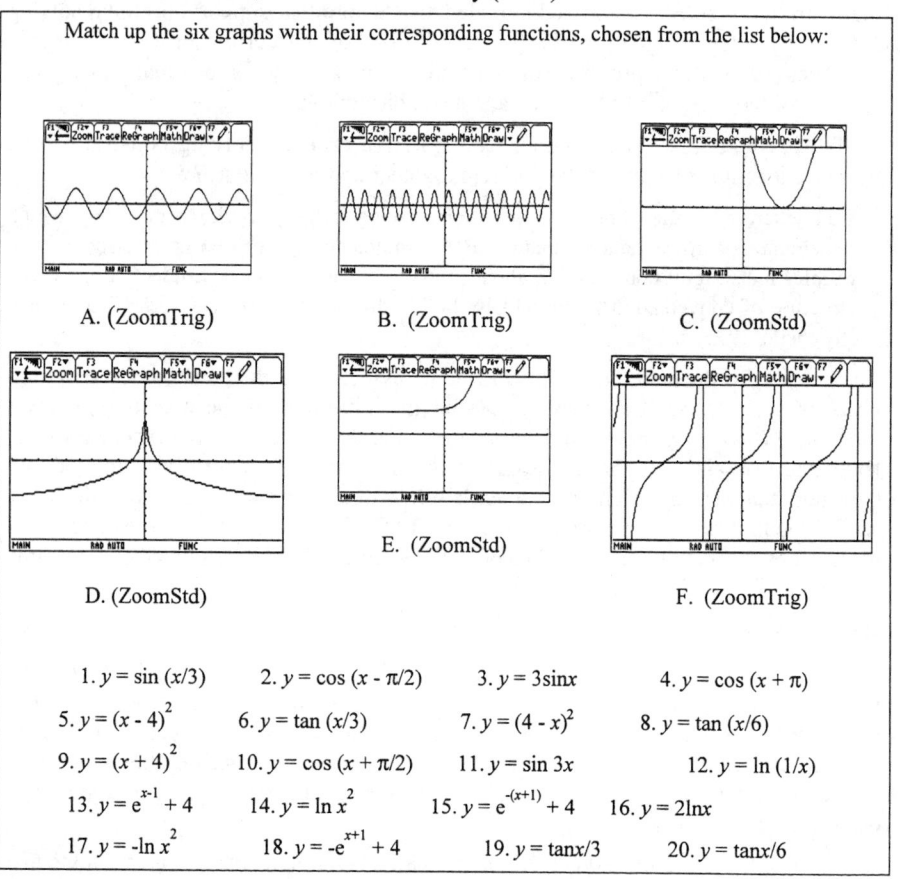

Match up the six graphs with their corresponding functions, chosen from the list below:

A. (ZoomTrig)   B. (ZoomTrig)   C. (ZoomStd)

D. (ZoomStd)   E. (ZoomStd)   F. (ZoomTrig)

1. $y = \sin(x/3)$ 2. $y = \cos(x - \pi/2)$ 3. $y = 3\sin x$ 4. $y = \cos(x + \pi)$

5. $y = (x - 4)^2$ 6. $y = \tan(x/3)$ 7. $y = (4 - x)^2$ 8. $y = \tan(x/6)$

9. $y = (x + 4)^2$ 10. $y = \cos(x + \pi/2)$ 11. $y = \sin 3x$ 12. $y = \ln(1/x)$

13. $y = e^{x-1} + 4$ 14. $y = \ln x^2$ 15. $y = e^{-(x+1)} + 4$ 16. $y = 2\ln x$

17. $y = -\ln x^2$ 18. $y = -e^{x+1} + 4$ 19. $y = \tan x/3$ 20. $y = \tan x/6$

Figure 1: Class Activity: Identifying the Graphs of Functions

The data examined in this paper pertains to a lesson where three GCE Advanced level Further Mathematics students, Robert, Martin and Julie [1] were asked to identify the symbolic forms of six graphed functions from a list of twenty possibilities and to discuss their ideas. These students were all experienced graphical calculator users and were each provided with a Texas Instrument TI-92 to assist them in their task. The task presented to the students is reproduced in Figure 1. The discussions surrounding three of the graphs are presented below.

DATA ANALYSIS

Notions developed by Teasley and Roschelle (1993) were used to analyse the interaction. Teasley and Roschelle propose that social interactions in the context of problem solving activity occur in relation to a *Joint Problem Space (JPS)*. They maintain that the *JPS* is a shared knowledge structure that supports problem solving activity by integrating (a) goals, (b) descriptions of the current problem state, (c) awareness of available problem solving actions, and (d) associations that relate goals, features of the current problem state and available actions.

In Teasley and Roschelle's model, collaborative problem solving consists of two concurrent activities: solving the problem together and building a *JPS*.

> Conversation in the context of problem solving activity is the process by which collaborators construct and maintain a *JPS*. Simultaneously, the *JPS* is the structure that enables meaningful conversation about problem solving to occur. Students can use the structure of conversation to continually build, monitor and repair a *JPS*. (Teasley and Roschelle, 1993, p. 236)

The analysis of the data thus involved finding evidence for the construction of a joint problem space as well as identifying student 'initiation' of the discourse, student 'acceptance' of arguments and cases of students 'repairing' misunderstandings. Evidence was also sought for instances that involved 'collaborative completions' between students, in which one student's turn would begin a sentence and the other student would use their turn to complete it. We shall present three examples of discussions to illustrate how the behaviour of individual students contributes towards the development of shared meanings.

Discussion of Graph A [$y = \cos (x - \pi/2)$]

The students' discussion of the possible symbolic representations for the first graph highlighted the way in which Robert's use of the graphical calculator provided a means through which he could become part of the *JPS* that was being created by Julie, Martin and the teacher-researcher. The discussion also signified the importance of the teacher's role in promoting collaboration and meaning making in a graphical calculator environment.

1   SE: Can anybody tell me which function represents the graph in the first one?

2	Martin:	Is it cos $(x + \pi/2)$?
3	SE:	And why do you say that?
4	Robert:	It's a sine graph.
5	SE:	Contradiction there. Explain your choice. [Directed at Martin].
6	Martin:	Er well it looks - it's got to be like sine or cos and I think that cos starts at the top and each line on the scale is 90° which is $\pi/2$ radians, so it's been moved ...
7	SE:	It's been moved across to the ...
8	Martin:	It's got to be $-\pi/2$ rads then because it's gone the other way, so it's cos $(x - \pi/2)$.
9	SE:	Ok so you think it's cos $(x - \pi/2)$. Why do you say that it might be a sine [graph]?
10	Robert:	Because sine of zero is zero and I'd say that that is in fact – because it seems that B is also a sine wave but that's more concentrated – I'd say that A is sin $x/3$.
11	SE:	You think that it's sin $(x/3)$?
12	Robert:	I wouldn't swear to it.
13	SE:	And what do you think? Have you got any ideas about this one?
14	Julie:	I think it's cos $(x - \pi/2)$.
15	SE:	And why do you think that it's cos $(x - \pi/2)$?
16	Julie:	It's been moved.
17	SE:	It's been moved?
18	Julie:	Yes it's a translation.
19	SE:	And in which direction is it moved?
20	Julie:	Er $\pi/2$ in the x-axis.
21	SE:	Yes. Ok so have you tried to actually graph on the TI-92 the first one that you thought it was?
22	Robert:	Yes.
23	SE:	And what did you get?
24	Martin:	Isn't that cheating drawing the graph to see which?
25	SE:	No, no he is just convincing himself.
26	Robert:	To be honest I can't remember what I typed in.
27	SE:	Well, let's think about the first one $y = \sin(x/3)$. What is the graph of that going to look like?
28	Robert:	Wide, and wider than it is there. [Robert pointed to graph A].
29	SE:	Yes. Ok, I'm going to say that you two are actually correct. Now it looks like a sine because it is sine of x, that is sin x.
30	Robert:	Yes.
31	SE:	But it can also be represented by $y = \cos(x - \pi/2)$ that's another...

32   Robert: I see where that's coming from.

When asked to identify graph A, Martin was the first to offer a suggestion: 'Is it cos $(x + \pi/2)$?' (line 2) in which he invited acceptance or repair. Robert on the other hand seemed more confident and asserted that this *is* a sine graph (line 4). Robert recognised the distinctive shape of the graph as being of the form $y = \sin x$ and as such did not initially think of the graph in terms of a translation of the cosine function, as Martin had suggested. When asked to explain his initial response Martin pictured the graph of cos$x$ in his mind and then considered the effect that the transformation cos $(x + \pi/2)$ would have. This allowed him to perform a self-repair, by recognising his original error and realising that the correct form of the given graph was actually cos $(x - \pi/2)$ (line 8).

In contrast, Robert who had immediately recognised the graph as that of sin x, was somewhat confused by the fact that this was not one of the listed options. His initial image of this function as a sine graph was strong and instead of considering the graph as a translation of cos x, he began to consider the other sine functions listed, focusing on sin $(x/3)$ (line 10). Yet, he was still uncertain that this was the correct function (line 12). At this point Julie who had remained silent throughout was drawn into the conversation (line 13). The teacher-researcher's questions encouraged Julie to elaborate on her initial explanation of why she had accepted Martin's argument and showed she had made sense of the problem. However, Robert seemed unaffected by the arguments proposed by Martin and Julie and in an attempt to clarify his thoughts he began using the graphical calculator.

When asked to consider what the graph of sin $(x/3)$ would look like in relation to graph A (line 27), Robert was able to recognise that the graph of sin $(x/3)$ would be wider than graph A. Use of the technology and the teacher-researcher's question aimed at making him think about the relationship between the two graphs had helped Robert to perform a self-repair. He now realised that sin $(x/3)$ was not the correct form of this function, and he started to question his initial thoughts and to eliminate the other sine functions listed. When it was explained that the graph could be represented symbolically by either $y = \cos (x - \pi/2)$ or $y = \sin x$, Robert remarked 'I see where that's coming from'. This suggested that he could visualise the action of the transformation f$(x-\pi/2)$ on the graph of f$(x) = \cos x$ and how this would produce the graph of sin x. He appeared to have internalised the argument that the teacher-researcher was presenting.

Thus, the use of the technology and the discussion in this example appeared to have resulted in some form of *cognitive reorganisation* for Robert. His thinking during the course of the episode had changed and by the end of this part of the discussion he was able to transfer his prior knowledge of trigonometric functions to this context. The concept of transformations became more meaningful to him, adding greater depth to his overall understanding of functions. He seemed to have begun to make the important visual connections between sine and cosine graphs and translations, which were also made explicit to Julie and Martin in this example.

The graphical calculator provided an authoritative means by which Robert could investigate the ideas being discussed and modify his own visual images of the graphs accordingly. Consequently, Robert's use of the technology was an important part of the process by which he was able to enter the *JPS*. Robert began to have more confidence in the arguments being posed by his peers following his graphical exploration with the technology. He was further convinced of the validity of these arguments through SE's concluding remarks. Here the teacher-researcher assumed the role of a more knowledgeable person in the Vygotskian zone of proximal development and helped all of the students to make sense of the apparent contradictions.

The use of the graphical calculator also provided a means of furthering the discussion and preventing a breakdown in communication. When Robert was unable to move forward he turned to the graphical calculator in an attempt to clarify his thoughts, rather than merely accepting the arguments put forward by Martin and Julie without really understanding them. The teacher-researcher's repeated questioning of the students' reasoning was also a factor in maintaining and steering the discourse and in the creation of a *JPS*. This allowed the students to make discoveries for themselves whilst receiving appropriate guidance and reinforcement of their solutions.

In this example the students were working in a group to find the solution to the problem. However, close examination of their dialogue indicates that they actually seemed to be working separately, within the group situation, towards achieving this common goal. There were plenty of interactions between each student and the teacher-researcher, but there was no direct interaction between the students themselves. Each of the students made their own independent contributions to the discussion, which appeared separate from previous utterances. They did take turns in the conversation, but did not question one another's contributions or request further elaboration. Robert did not seem to consider the contributions made by Martin and Julie. Martin and Julie did not question Robert's argument. Julie did not make clear her agreement with Martin's line of reasoning until she was specifically asked to share her viewpoint. Each student appeared to take his or her turn in the conversation in a linear way. There were no interruptions from the other two students when one of them was presenting their argument.

The way in which these students were interacting made it necessary for the teacher-researcher to take an active role in maintaining and encouraging the discussion between students, in verifying students' assertions and in providing clarification and explanation of solutions where needed. There was also a need to promote the use of the technology, especially in Martin's case, as he saw this as a means of cheating (line 24) rather than as a tool that could help the students to clarify their thoughts and move towards a different level of understanding.

At the end of the episode the students had gained some shared understanding of the problem posed and of the solution through the creation of a *JPS*. They had each participated in the same activity at the same time and they had all listened to each other's contributions and those made by the teacher-researcher. The knowledge and understanding that these students had gained as a group appeared to be greater than that which they already possessed as individuals and as Jones and Mercer (1993) argue, this indicates that successful learning had taken place.

Discussion of Graph B [$y = \sin 3x$]

The students' discussion of graph B showed that successful collaboration can occur in a graphical calculator environment without the use of this technology. However, we propose that even when the graphical calculator was not being used, as is the case here, it was still having an effect on the students' thinking.

1	SE:	Can anybody think of a function for B?
2	Martin:	I reckon its sin 3x.
3	SE:	Sin 3x.
4	All:	Yes.
5	SE:	You seem to agree on that one. So how did you come up with that conclusion?
6	Robert:	There doesn't seem to be any sneaky cosine tricks.
7	SE:	Not this time.
8	Martin:	It's a sine wave and it's been er...
9	Robert:	Three times x would condense it.
10	Martin:	It's got a stretch parallel to the x-axis of a third, because it got closer together.
11	SE:	Yes, you're all right it's sin 3x.

Martin initiated the discussion by asserting that this was the graph of sin 3x (line 2) and this time appeared to be much more confident with his suggestion. The other two students immediately accepted that this was the correct form of the function and when asked to give reasons why, both Martin and Robert took turns to construct an explanation, each building and elaborating on the previous utterances, thereby producing a collaborative completion (lines 8, 9, 10). When Martin paused to think (line 8), Robert anticipated what he may have intended to say and completed his statement. Together they provided a convincing argument for their choice of function. The students were thus all confident that they had identified the function correctly. Although, Julie did not participate verbally in this part of the discussion, she did make gestures that indicated her agreement with the arguments being put forward.

The knowledge constructed by the students in this example appears to be shared between them, especially Martin and Robert, and these two students appear to have constructed a *JPS*. However, it is uncertain whether Julie had actually developed a fully shared sense of the solution to the problem, as she did not verbalise her ideas through interaction with the other students. Martin and Robert seemed to be more supportive of each other when discussing this graph than when discussing the previous one. Rather than concentrating on developing their own arguments separately, they produced a joint explanation of why sin 3x was the symbolic representation of graph B.

Martin and Robert were able to perform a collaborative completion together because it seems their visual images of the function were strong and corresponded to one another. In this case each student appeared to be able to visualise clearly the effects of the transformation, without using the technology. Yet, there was evidence that the graphical calculator was having an impact on the way in which these students were thinking about the problem. In particular, Robert was now actively looking for alternative symbolic forms for the graphed functions following his exploration of the function represented by graph A with the graphical calculator (line 6). We therefore propose that the use of the graphical calculator and its continued presence in the environment restructure the way in which students think about problems, and that this is most productive when used as part of a local community of practice. This will be discussed further below.

Discussion of Graph F [$y = \tan(x/3)$]

The discussion of the final graph serves to illustrate how collective use of the graphical calculators enabled the students to establish a *JPS*.

1	Robert:	It's a tangent.
2	SE:	Think about the scale the TI-92 uses.
3	Robert:	To see if it was increasing, I could just draw the normal graph.
4	SE:	Ok, if it helps you can draw the – you can all draw the tan x graph and see what happens on your machine and then from there you can hopefully deduce what the function is.
5	Robert:	It's a stretch of factor 3.
6	Martin:	It's tan of x over 3.
7	Robert:	Yes.
8	SE:	Is that $y = \tan(x/3)$ or $y = (\tan x)/3$ because there are two of them?
9	Martin:	$y = \tan(x/3)$.
10	SE:	$y = \tan(x/3)$ and what do you think? Have you managed to get the tan?
11	Julie:	Yes. That's the whole thing. [Julie pointed to the tan x in tan x/3].
12	SE:	That's tan of x all divided by 3.
13	Julie:	So yes $y = \tan(x/3)$.
14	SE:	$y = \tan(x/3)$, yes well done you are right.

Robert was the first to state that this graph belonged to the tangent family of functions (line 1). There was, however, some uncertainty amongst the students as to what the graph of $y = \tan x$ would look like in relation to graph F. This was evident in the silence that followed Robert's initial suggestion. Recognising this problem, the teacher-researcher asked the students to think about the scale that the graphical calculator uses to draw trigonometric functions (line 2). Robert then suggested that he could draw the graph of tan x using the TI-92 and compare this with graph F to deduce the relationship (line 3). The teacher-researcher then advised all three students to try this approach. Robert compared the graphs and deduced that graph F was obtained using a stretch of factor three. To complete Robert's statement, Martin added that the correct function was "tan of x over three" and Robert immediately agreed. As in the discussion of graph B, Robert and Martin attempted to construct shared knowledge and again produced a collaborative completion. However, as there were two functions which could be verbalised as "tan of x over three", the teacher-researcher sought confirmation that Martin had identified the function correctly and was quickly satisfied that he had. Up until this point Julie had not contributed to the discussion and the teacher-researcher drew her into the conversation again to see if she was following the arguments being presented. Julie accepted the choice of function offered by Martin and provided some evidence that she had understood why this was the correct function (line 11), which was then confirmed by the teacher-researcher's closing comments.

The students had thus been able to develop some shared understanding of the transformations used in this example through the creation of a *JPS*. Moreover, the use of the graphical calculator was an important part of this process. The interaction between the students was constructed in relation to the graphs produced by the graphical calculator and it was this factor that led directly to the collaborative completion between Martin and Robert. This occurred because the students were able to establish a shared visual interpretation of the function using the graphical calculator. In other words, in this episode use of the technology provided Julie, Martin and Robert with a common starting point from which they were able to think about the problem in the same visual terms. From this position they were each able to contribute towards correctly identifying the symbolic form of the function. As their discussion was structured around their shared use of the graphical calculator, this facilitated successful interaction. The decision to use the graphical calculator in this case thus proved extremely productive and resulted in the collaborative completion.

Local Communities of Practice in Action

During these discussions a *local community of practice* appears to have been established by the students and the teacher-researcher. Figure 2 summarises the kinds of interactions used by each participant throughout the discussion of all six graphs.

Firstly, as shown by Figure 2, the students each showed willingness to explore and explain ideas to one another. They clearly saw themselves as functioning mathematically within the lesson, as they were each offering suggestions as to which functions represented the given graphs, based on some mathematical reasoning, which enabled them to obtain the correct form of the function in each case.

Secondly, the teacher-researcher ensured that the students received public recognition of their competence. This was achieved through acceptance of the students' ideas ("yes, well done you are right", "yes, you are all right, it's sin $3x$").

Thirdly and most significantly, as the discussions progressed, the students began actively working together towards achieving a common understanding of each problem, through the sharing of ideas and questioning of one another. This led to the creation of *joint problem spaces* and successful collaboration in the form of collaborative completions between the students themselves and with the teacher-researcher.

Fourthly, the students each shared behavioural traits, such as presenting and justifying their own arguments and listening to, accepting and questioning the arguments of others. The language used by the students was both scientific and natural, and the students appeared to have shared conceptions of the scientific language that was used. The students also used the graphical calculator together as a group in their attempts to identify the fifth graph (not presented here).

	Robert	Julie	Martin	SE
Presenting ideas	4	2	1	1
Explaining ideas	3	3	0	3
Making assertions	7	2	3	1
Making statements	6	1	2	5
Showing acceptance	4	4	3	6
Repairing ideas	2	0	0	3
Self repairing ideas	1	0	1	0
Questioning	1	2	2	23
Performing collaborative completions	4	0	2	2
Number of interactions involving natural language	18	9	8	25
Number of interactions involving scientific language	14	5	6	19
Total number of interactions	**32**	**14**	**14**	**44**

Figure 2: Types of interaction used by the participants

The shape of the lesson depended on the active participation of the students. Each student created his or her own role in the practice, which varied accordingly. During the discussions the patterns of interactions between the students were continually changing as each new graph was considered. In each case the individual students appeared to occupy different positions within the discussion, modifying their roles depending on their needs. Martin initiated the discussion around the first two graphs, and Robert took over this role for the discussions concerning the remaining four graphs. Robert also began to act as a more capable peer in the Vygotskian *zone of proximal development* (graphs C-F). He continually made verbal contributions to the discussions and at times took control of the discussion, while Julie and Martin spent more time actively listening and thinking rather than speaking. In most cases Julie did not contribute voluntarily to the discussions and needed to be drawn into the discourse.

Martin was initially quite instrumental in moving the group towards the correct solutions (graphs A and B). However, as Robert took over initiation and steering of the discourse, Martin seemed to fade into the background. He was unsure about the symbolic forms of some of the graphs and he made fewer verbal contributions when discussing these, especially graph E (see Elliott, 1999). He needed time to take into full consideration the arguments offered, to enable him to form his own ideas and to convince himself of their meaning. In this way Martin was attempting to derive his own meaning from that which already belonged to Robert and Julie.

Julie operated as an active listener during the majority of the discussion, only offering her suggestions when specifically asked to do so. It was only during the discussion of graph E that this pattern changed and her contributions became more spontaneous. Her difficulties with this graph encouraged her to share her ideas more freely, in an attempt to derive meaning from the interaction. However, at all other times she was a participant with a relatively voiceless role in the *local community of practice*.

Finally, both the students and the teacher-researcher regarded themselves as being involved in the same activity. In each case the teacher-researcher attempted to initiate each student into the discourse with the aim of encouraging the construction of shared knowledge. Through these actions the teacher-researcher attempted to maintain and repair the *JPS* that was being created. Figure 2 highlights that questioning formed an extremely important part of the teacher-researcher's strategy for encouraging participation and the construction and maintenance of a *JPS* amongst the students and herself. It also illustrates the quality and frequency of interactions made by Robert in comparison to Martin and Julie. Robert performed more successfully overall in the class activities in this trial, taking into account the scores that he obtained in the entire series of exercises, too numerous to present here, which comprised this investigation as a whole. This may have been the result of his additional willingness to share ideas and difficulties with the other students and the teacher-researcher and to use the technology without being prompted to do so.

This type of environment, where *local communities of practice* are established in which the students and teacher are able to construct and maintain *joint problem spaces*, is seen to be conducive to successful collaborative work involving graphical calculators. Within this supportive environment students are able to establish an effective means of operating with graphical calculators, in which the knowledge generated is shared amongst the participants. As seen in the discussion of graph B, this may not necessarily involve working with graphical calculators all of the time. For example, if the students are able to visualise the effects of a particular transformation on the graph of a function effectively without the aid of technology, then they may choose not to use the graphical calculator in that instant. However, the way in which they approach the problem is likely to reflect their prior use of the technology.

REFLECTIONS

The use of the graphical calculator in this study enabled the students to perform collaborative completions together and thereby created the opportunity for effective collaboration. This was achieved because the students were able to produce a shared visual representation of the problem using the graphical calculator and thereby create *joint problem spaces*. Successful collaboration was also promoted by the ability for students to reinforce their arguments through use of the graphical calculator. This discouraged breakdowns in communication between the students. The ability of students to use the graphical calculator as a means of verifying or, in particular, disproving their ideas also led to *cognitive reorganisation*, especially in Robert. This enabled Robert to gain a better understanding of the actions of transformations and the corresponding relationships between the graphs of sine and cosine functions. Input from the teacher-researcher was also seen to be a contributing factor in this cognitive reorganisation process.

There is an important role for the teacher in initiating the students into the *local community of practice* through initiating discussion of the results obtained by the graphical calculator. This could be on a one-to-one basis, in small groups or as a whole class. Through the interaction with the teacher, peers and the technology the individual student is able to develop a more meaningful understanding than was hitherto possessed. Analysis of the episodes has pointed to the centrality of the teacher's role in maintaining and encouraging discussion between the students, especially in relation to the results produced by the graphical calculators, and in providing additional verification of these results and the students' assertions. A further important function of the teacher lay in providing clarity and explanation of the results of the students' exploration with the technology, especially when the students were unable to reach a common understanding of their findings by themselves. The teacher needs to scaffold the students' learning with the graphical calculators to ensure that the technology is used effectively and results are interpreted correctly by the students, so that any misunderstandings are not perpetuated.

The establishment of *local communities of practices* in the classroom was seen to be conducive to successful collaborative work involving graphical calculators. In this type of supportive environment the students shared ownership of their use of the technology and they and the teacher-researcher could build and maintain *joint problem spaces*, which can lead to graphical calculators being used to greatest effect. It was not essential for the students to use graphical calculators all of the time in order for learning to be successful in their community of practice. Yet, the way in which the students operated whilst using the graphical calculators was seen to influence the way in which they approached problems without use of the technology.

The ways in which students define their eventual roles in a *local community of practice* is a complex process. The patterns of interactions between the students changed as each new graph was considered. Throughout the discussions the individual students appeared to occupy different positions within the discourse, modifying their roles depending on their needs. Robert, in particular, adopted the role of initiating and steering the discussions, whilst reacting to the arguments presented by the other students. So as the discussion developed, Robert's positioning within the discourse evolved and he proceeded to occupy a central role. He was an eager and very active participant and his position developed into that of a more capable peer in the *zone of proximal development*.

None of the students appeared to be self-reliant. They learned from each other, from the teacher-researcher's comments, from using the technology and by participating in the community of practice, through their discourse. The positions that the students occupied within the discourse were their ways of appropriating meaning and contributed towards their success as learners. Moreover, the successful learning that took place in these discussions is attributed to the creation of *joint problem spaces*, which can be thought of as particular examples of *local communities of practice* in action and both the technology and the teacher-researcher played an important part in contributing towards their development.

NOTE

1. Throughout this paper the teacher-researcher is SE and pseudonyms are used for students.

REFERENCES

Berger, M.: 1998, 'Graphical calculators: an interpretative framework.' *For the Learning of Mathematics, 18*(2), 13-20.

Borba, M. C.: 1996, 'Graphing calculators, functions and reorganisation of the classroom.' *The Role of Technology in the Mathematics Classroom. Proceedings of Working Group 16 at ICME-8, the 8th International Congress on Mathematics Education* (pp. 53-60). Seville: Spain.

Eisenhart, M. A.: 1988, 'The Ethnographic Research Tradition and Mathematics Education Research.' *Journal for Research in Mathematics Education, 19*(2), 99-114.

Elliott, S.: 1998, 'Visualisation and using technology in A level mathematics.' In E. Bills (ed.), *Proceedings of the BSRLM Day Conference at the University of Birmingham* (pp. 45-50). Coventry: University of Warwick.

Elliott, S.: 1999, 'How does the way in which individual students behave affect the shared construction of meaning?' In E. Bills (ed.), *Proceedings of the BSRLM Day Conference at the University of Warwick* (pp. 13-18). Coventry: University of Warwick.

Elliott, S., Hudson, B. and O'Reilly, D.: 2000, 'Visualisation and the influence of technology in 'A' level mathematics: a classroom investigation.' In T. Rowland and C. Morgan (eds.), *Research in Mathematics Education Volume 2. Papers of the British Society for Research into Learning Mathematics* (pp. 151-168). London: British Society for Research into the Learning of Mathematics.

Jones, A. and Mercer, N.: 1993, 'Theories of learning and information technology.' In P. Scrimshaw (ed.), *Language, Classrooms and Computers* (pp. 11- 26). London: Routledge.

Lerman, S.: 1994, 'Metaphors for mind and metaphors for teaching and learning mathematics.' In J. Ponte and J. Matos (eds.), *Proceedings of the 18th Conference of the International Group for the Psychology of Mathematics Education* (Vol. 3, pp. 144-151). Lisbon, Portugal.

Lerman, S.: 1996, 'Intersubjectivity in mathematics learning: a challenge to the radical constructivist paradigm?' *Journal for Research in Mathematics Education, 27*(2), 133-150.

Moll, L. C.: 1990, 'Introduction.' In L. C. Moll (ed.), *Vygotsky and Education: Instructional Implications and Applications of Sociohistorical Psychology* (pp. 1-27). Cambridge: Cambridge University Press.

Pea, R. D.: 1987, 'Cognitive technologies for mathematics education.' In A. H. Schoenfeld (ed.), *Cognitive Science and Mathematics Education* (pp. 89-122). Hillsdale, NJ: Lawrence Erlbaum Associates.

Teasley, S. D. and Roschelle. J.: 1993, 'Constructing a joint problem space: the computer as a tool for sharing knowledge.' In S. P. Lajoie and S. J. Derry (eds.), *Computers as Cognitive Tools* (pp.229-257). Hillsdale, NJ: Lawrence Erlbaum Associates.

Winbourne, P. and Watson, A.: 1998, 'Participating in learning mathematics through shared local practices.' In A. Olivier and K. Newstead (eds.), *Proceedings of the 22nd Conference of the International Group for the Psychology of Mathematics Education* (Vol. 4, pp. 177-184). Stellenbosch, South Africa.

Vygotsky, L. S.: 1978, Mind in Society. The development of higher psychological processes. Cambridge, MA: Harvard University Press.

14 CONJECTURING IN OPEN GEOMETRIC SITUATIONS USING DYNAMIC GEOMETRY: AN EXPLORATORY CLASSROOM EXPERIMENT

Federica Olivero

Graduate School of Education, University of Bristol

In this paper I describe a classroom teaching experiment carried out with a class of Year 10 students. This experiment was twofold. On the one hand it was aimed at developing and trying out a new mode of working in the classroom, taking into account the possibilities offered by dynamic geometry software as support in the conjecturing and proving process in geometry. On the other hand, from the research point of view, it provided the possibility of testing out in the classroom a theoretical model describing and interpreting students' use of Cabri-Géomètre in open geometric problems, with a particular focus on dragging. The main findings relate to the evolution in the use of dragging in Cabri and the production of rich conjectures, which can provide the basis for development and evolution towards the proving process.

INTRODUCTION

The classroom teaching experiment [1] discussed in this paper is the first phase of a project [2], aimed at studying how students can be supported to accept and understand the role of proof in school mathematics, through engaging with a computer-based microworld, namely the software Cabri-Géomètre (Baulac, Bellemain and Laborde, 1989). In particular the study investigates the process of transition from explorations and conjectures to proving in geometry, as it can be supported by a dynamic geometry environment. This paper focuses in particular on the transition from exploration to conjectures, as a basis for the transition to proofs.

The major project falls within the framework of "research for innovation" (Arzarello and Bartolini Bussi, 1998), whose aims are: on the one hand to produce innovations with respect to content and methodology in the classroom, and on the other hand to develop theoretical models based on classroom observations, which may help teachers better understand students' learning processes and may then effectively be used by teachers in the classroom. In this respect, researchers and teachers work collaboratively in all the phases of the research project (Arzarello and Bartolini Bussi, 1998). In this way the research and analysis is grounded in practice, which improves its validity and the usefulness of its insights to other practitioners.

According to this perspective, the purpose of the small-scale experiment described in this paper was two-fold. On the one hand it was aimed at developing and trying out a new mode of working in the classroom, taking into account the possibilities offered by new technologies in the learning of mathematics, with particular respect to dynamic geometry software as support in the conjecturing and proving process in geometry. On the other hand, from the research point of view, it provided the possibility of testing out in the classroom a theoretical model describing and interpreting students' use of Cabri in open geometric problems. At the same time this classroom experiment was useful in giving insight into the model itself and suggesting further development and refinement.

Taking into account these two directions, the paper is structured into three parts. The first part provides a brief account of the theoretical background on which the classroom experiment was grounded and the methodology of research that has been used. The second part is an overview of how the classroom experiment was planned and carried out in the classroom. The final part presents the analysis of some students' productions with respect to their use of Cabri and the production of conjectures. Preliminary conclusions and issues for further discussion are introduced.

THEORETICAL BACKGROUND

The work described here sits within the current discussion in mathematics education about the teaching and learning of proof in school mathematics and the use of dynamic geometry software in the mathematics class. One characteristic of the debate around proof amongst researchers in mathematics and mathematics education (e.g. Balacheff, 1998; Duval, 1991; Hanna, 1996; Mariotti, Bartolini Bussi, Boero, Ferri and Garuti, 1997) concerns the role of explorations and conjectures with respect to proving and proof. In the literature, there is a wide range of opinions concerning this issue. Some authors (Arzarello, Gallino, Micheletti, Olivero, Paola and Robutti, 1998; Boero, Garuti and Mariotti, 1996) say that exploration and conjecture is a key phase in the overall cognitive process of proving. It is possible that the arguments needed to prove a statement develop within this initial phase, before being re-organised in a deductive way in order to form conditional sentences. My project investigates this perspective.

A considerable body of research has also pointed out the problems and difficulties in the teaching and learning of proof in school mathematics (Chazan, 1993; Harel and Sowder, 1996; Healy and Hoyles, 1998). Usually students do not understand the meaning of the act when they are asked to prove a theorem (Chazan, 1993). Moreover most of the time students are presented with ready-made proofs, which they do not understand (Usiskin, 1982). In some countries proof is disappearing from the curriculum (Hanna, 1996); for example in the UK, proof has become inaccessible to the majority, in that it is only at Level 7 or 8 of AT1 [3], that students are expected to prove their conjectures in any formal sense (Hoyles, 1997).

New issues about the role of proof arise when new technologies are taken into account and dynamic geometry software such as Cabri-Géomètre is introduced into school practice (Hanna, 1996). Cabri is not a tutorial system, therefore it cannot be autonomous from the learning point of view, that is, it cannot be considered responsible on its own for the devolution and acquisition of mathematical content. As a consequence, educators must avoid the 'fingertip effect' (Perkins, 1993), that is, simply making a support system available and expecting that people will more or less automatically take advantage of the opportunities that it affords. On the contrary Cabri should be seen as a 'conceptual reorganiser' (Pea, 1987), in that it provides users with new tools that were not available when carrying out paper and pencil geometry. For instance, introducing Cabri provides the learner with two worlds (Sutherland and Balacheff, 1999): a theoretical world, which is that of geometry, and a mechanical, manipulative world, which is the phenomenological domain of Cabri. The possibility of dragging things around the screen is one of the most important affordances of Cabri. As Goldenberg (1995, p.220) observes: "Line segments that stretch and points that move relative to each other are not trivially the same objects that one treats in the familiar synthetic geometry, and this suggests new styles of reasoning"; this is 'dynamic geometry'.

Research is currently being carried out about the potentialities of dragging (Goldenberg and Cuoco, 1998; Laborde, 1995; 1998; Mariotti, 1998). In particular dragging is studied with respect to the support it can provide in the conjecturing and proving process (Arzarello *et al.*, 1998): a classification of different dragging modalities students might use in solving a problem in Cabri has been produced and the cognitive counterpart of dragging has been investigated (Arzarello *et al.*, 1998; Arzarello, Olivero, Paola and Robutti, 1999; Olivero, 1999). In fact, classroom observations showed that students' use of dragging changes according to the different aims they want to achieve: exploring a situation; trying to make conjectures; validating conjectures; testing conjectures; attempting to construct a justification. The classification of dragging modalities is presented later in the paper.

As far as the mode of working in the classroom is concerned, this experiment draws on different models. Arsac, Germain and Mante (1988) present the use of open problems in the classroom, trying to characterise a teaching and learning activity which allows students to 'do' mathematics. Open problems are characterised by development over four phases: exploring-conjecturing-validating-proving. Students are faced with a situation in which there are no precise step-by-step instructions, but they are left free to explore and make their own conclusions. With particular respect to geometry, the structure of open problems can be characterised as follows (Mogetta, Olivero and Jones, 1999; Olivero, 1998):

• the statement is short, and does not suggest any particular solution method or the solution itself. It usually consists of a simple description of a configuration and a generic request for a statement about relationships between elements of the configuration or properties of the configuration.

• the questions are expressed in the form "which configuration does...assume when...?" "which relationship can you find between...?" "What kind of figure can...be transformed into?". These requests are different from traditional closed expressions such as "prove that...", which present students with an already established result.

When using open problems, the role of the teacher becomes very important, not only in the preparation of the situation but also in the "classroom discussion" (Bartolini Bussi, 1998), which is a discussion of students' discoveries and solutions to a given situation, orchestrated by the teacher, through which validated results and processes should emerge.

METHODOLOGY

The research involved two researchers (including the author), a teacher and his classroom. It followed different stages: choosing a topic to teach; developing a sequence of activities both in Cabri and with paper and pencil in the classroom; developing worksheets; carrying out the activities with students; analysing students' productions. Since the classroom intervention involved an innovative mode of working in the classroom, both for the teacher and for the students, the planning of the whole sequence was important. Moreover just providing students with the software Cabri would not have provoked the intended learning ("fingertip effect", Perkins, 1993).

In this particular case the intervention and the learning objectives were agreed with the teacher, who was willing to start using computers in his mathematics classroom, but then I took much responsibility for the planning and carrying out of the experiment as the teacher had never worked with computers or with the kind of open activities that were used. This allowed the teacher to gain confidence in this new environment and to 'see' that doing something different in the classroom is really possible. He said he would not have been able to do such a thing on his own, therefore he found the researchers' help very valuable; next year he will probably repeat the same thing, and now on his own.

During the work in the classroom, the two researchers (including the author) were participating together with the teacher; we acted as participant observers, both helping the teacher to handle the situation and trying to keep records of the general classroom situation and of the students' work. The students had been told that we came from the University and we were presented as people who would help them and the teacher in some sessions. The data gathered from the experiment were: field notes of each session; all written work by students; audiotapes of one pair of students for each session.

DESCRIPTION OF THE CLASSROOM EXPERIMENT

The students. The intervention was carried out in a class of 25 Year 10 (15-year-old), mid-top set male students. This is a traditional class, in which students are typically taught with the teacher teaching the whole class from the front. They were not used to working in groups. The problems they usually have to deal with in the mathematics class are textbook questions that are mainly straightforward questions requiring immediate answers.

"We rarely do problems, just questions" [4]; "In the classroom the problems are shut; in Cabri everything is more open-ended."

They had never been asked to make conjectures or give justifications.

Aims of the intervention. The general learning objectives addressed by the intervention were: using Cabri as a tool for discovering and conjecturing; making explorations and conjectures; validating conjectures; justifying conjectures. We didn't want students to become Cabri experts, but to get the idea of the possibility of doing mathematics differently with dynamic geometry.

The topic chosen in negotiation with the teacher was 'circle theorems' [5]. The teaching sequence was aimed in particular at supporting students to discover circle theorems by working them out themselves in Cabri, instead of telling them the established results. Reconstructing the process of discovery of a theorem might be a way towards the construction of its justification (Arzarello *et al.*, 1998).

Sequence in the classroom. The sequence of prepared activities aimed at introducing Cabri to students who had never used it before, and then at leading them towards increasingly more complex geometrical problems around circle theorems, involving a basis for proving. The activities centred around open problems, in which students were asked to explore a situation, make conjectures, validate them and try to justify them. The structure of the activities was changed during the development of the sequence, according to students' responses. Some of them found the activities too open and the wording not clear:

"they were very vague, I couldn't understand some."

At the beginning they found it difficult to get started, they had to be guided through more specific questions. Some others liked the open structure of the problems very much:

"the way they required you to think and design on the computer gave a lot of freedom"; "make you think and give a lot of answers."

In the computer room students worked in pairs at the computer, following a particular problem presented on paper. At the end of each session a summary of what had been going on was carried out by either the teacher or one of the researchers. The sessions in the classroom consisted both of group work and classroom discussions. Even if students had never worked in groups before, working in groups turned out to be very powerful with respect to their productions. The students themselves stressed the importance of the interaction with other people in such an open activity. First of all, it was useful as a way of getting and giving help to each other:

"one doesn't know, the other can help."

Secondly, they liked the fact they could discuss and compare different ideas:

"we can express our ideas and hear others' ideas."

Working in groups encourages students to make explicit what they think through talking; in this sense it can help the formalisation of what they see in Cabri, in that they need to make explicit the relationships between the objects which move in Cabri. Moreover this is a powerful research tool, as it provides a window on students' cognitive processes.

The experience lasted two months. Each session lasted 40 minutes. We had nearly one session per week. Sessions 1, 3, 4, 6, 8, 9 took place in the computer room, while sessions 2, 5, 7, 10 took place in the mathematics classroom.

Session 1 gave students the possibility of exploring what Cabri is without any precise instruction. *Session 2* consisted of a test on proof (Healy and Hoyles, 1998) aimed at developing an idea of students' conceptions and knowledge about proof. *Session 3* was devoted to a more structured exploration of Cabri, with the purpose of supporting students to learn the meaning of constructions in Cabri and a way to check the correctness of their constructions, i.e. the dragging test. In *Session 4* students had to make conjectures in Cabri by using dragging in order to explore and discover properties. *Session 5* consisted of a classroom discussion in which students had the possibility of sharing their discoveries with their classmates. *Session 6* was the first activity aimed at introducing one of the circle theorems (1.). *Session 7* was a follow-up lesson, in which a general discussion about what students found and proved was orchestrated by a researcher (the author). *Sessions 8 and 9* had the same structure as session 6 and were aimed at introducing theorems 2 and 3. *Session 10* was devoted to constructing a proof for one theorem; an evaluation questionnaire was handed out in order to obtain a feedback of the whole intervention from both the students and the teacher. The worksheets used in the classroom are in the Appendix.

THE USE OF CABRI AND CONJECTURING: AN ANALYSIS OF STUDENTS' PRODUCTIONS

Drawing on the perspective that conjecturing is a key phase in the proving process and that dragging seems to be one of the most powerful features of Cabri from the didactical point of view, this section will focus on the analysis of some students' work during the conjecturing phase, with respect to the way they used Cabri. In particular the different uses of dragging will be analysed, according to the framework described in Arzarello *et al.* (1999) and Olivero (1999); the influence of the dragging modalities over the form of the conjectures produced by students is also mentioned.

The main dragging modalities which have been classified are the followings. *Wandering dragging*, that is moving the basic points on the screen randomly, without a plan, in order to discover interesting configurations or regularities in the figures. *Guided dragging*, that is dragging the basic points of a figure in order to give it a particular shape. *Soft locus dragging* [6], that is moving a basic point so that the figure keeps a discovered property; the point which is moved follows a path (*soft locus*), even if the users do not realise this (an example is presented later). *Dragging test*, that is moving draggable or semi-draggable [7] points in order to see whether the figure keeps the initial properties: if so, then the figure passes the test; if not, then the figure was not constructed according to the geometric properties you wanted it to have.

It was observed that students exploited these different dragging modalities in order to achieve different aims, such as exploring, conjecturing, validating, justifying. For example, *wandering* and *guided dragging* were generally used in the discovering phase, *soft locus dragging* marked the construction of a conjecture, *dragging test* was mainly used to test a conjecture. Therefore looking at how students used dragging provided an insight into their cognitive processes.

In this classroom experiment two different levels need to be considered: the overall classroom experience over the sessions and the single activity of dealing with a problem in Cabri in one session, centred around exploration and production of conjectures (and eventually justifications). In general an evolution occurred at both levels: a long-term evolution in the use of Cabri within the whole intervention and an evolution in terms of dragging modalities within each session.

The students' work showed an evolution towards exploiting all the potentialities of the dragging function: they moved from drawing only to constructing figures and exploring the richness of the situation proposed. In the first session students produced colourful pictures, which for them had the simple status of drawings. The second Cabri session was more productive in terms of students' appropriation of the Cabri features. I could see that almost all the students understood the use of the dragging test, as they then used it later on. In the evaluation questionnaire they wrote what the dragging test is:

> "if you can drag it and it stays mathematically similar then the test is positive", "used to prove a conjecture to see if it is true for other positions."

These answers show the comprehension of the dragging test both as a test in the construction of a figure and as a validation for a conjecture. In general the students found it rather more difficult to use dragging in order to explore a situation. At the beginning they did not move many things, but just looked at the figures on the screen. When constructing a figure they tried to put it in a 'stereotypic' position (for example, with the sides parallel to the border of the screen), without considering the fact that they could move the figure around after construction and therefore the starting position is not important. After some time, starting from the third Cabri session, they started to move things more but still without seeing many things. Most of the students moved points very fast and without any particular intention; in this way they could only see general things they missed relationships between points and specific changes in the figures. For example, in session 4, students moved the triangle randomly and fast so that they did not pay attention to the relation between the triangle and its circumcentre (outside or inside the triangle). As they continued, they began to exploit all the potentialities of dragging to make their conjectures, as can be observed in the examples below.

Evolution in the solution process of one Cabri activity will now be considered and two groups' production in two different sessions will be analysed with respect to the analytical model concerned with dragging modalities previously described.

The following is the solution process of one group for the problem *Angles in circle 1* [8] in session 6.

Students' work (students S and O)	Analysis of the use of dragging and the solution process
They obtain a right-angled triangle with CB horizontal (Figure 1). **Figure 1**	At first they draw a triangle in Cabri, then they use dragging in order to obtain a triangle with the required property. The only one they can find is the 'typical' one, that is with the sides parallel to the borders of the screen. In this first attempt they use Cabri in order to draw what they had in mind, like on paper. It is a kind of *guided dragging*.
S: Here's one… O: How many can you have? S: Just one…no two. They move C across AB to get to the other side (Figure 2). **Figure 2**	The fact that they can move C allows them to see another triangle with the required property. But again dragging follows what students had in mind beforehand, in that they say first what they think and then they do it in Cabri. They are not yet using Cabri for discovering conjectures. The kind of reasoning they carry out is: there is this one (Figure 1), but since we can drag C, there will be also the other one (Figure 2). And then they do it in Cabri.
S: There is two…no four, one for each vertex of the square (Figure 3) **Figure 3**	The same reasoning leads them to say there are four.

O: Try to move C.... How many triangles can you get?... S: You go on a circle. They move C trying to keep the angle 90°. S and O: You can have infinite triangles! And then they write down their conjecture: "If a circle is *drawn*, with its centre on the midpoint of AB. Then if *C is moved* to the perimeter of the circle then the angle ACB will be 90°" (Figure 4). 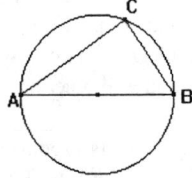 **Figure 4**	Now they exploit dragging in order to discover a conjecture. They move C looking at the measure of the angle. At first they do very small movements, not to lose 90°. Then they go a bit further. They drag C in order to keep the property they want (*soft locus dragging*) and they realise they are moving on a circle, without actually seeing it. Therefore now they read what happens on the screen. So they come to say there are infinite triangles. At this point they make their conjecture, translating what they see in Cabri (C moving on a 'imagined' circle) in a logical form (if...then). Their conjecture is still linked to the Cabri-world, as they refer to the movement of points; however they are able to provide and read the logical dependence of dragged points.
After writing down I ask them to check their conjecture in Cabri. They draw the circle with diameter AB, they put C on the circle. S: Yes, it's true.	After that, they are prompted to use the *dragging test* in order to check their conjecture. They know how to do it and probably for them this provides the required validation of the conjecture.

This production is representative of the way most students approached the problem. In some other cases, it was observed that the exploration at the beginning was quite random while then it increasingly focused on investigating the path of either A, B or C when the angle C was kept a right-angle.

The following is the production of another group concerning the problem *Angles in circle 2* [9] in session 8, which shows a different approach.

Students' work	Analysis of the use of dragging and the solution process
They draw a triangle ABC. Then they make the angle C 60°. They conjecture that the locus of C as A and B were fixed would be a circle and that A and B would not be on the circle.	At the beginning these students do not use Cabri to conjecture. They are trying to read this problem adapting the knowledge they have from the previous session. So they make a conjecture beforehand.

They try to use 'trace' [10] to test their conjecture. So they move the point C trying to keep the angle 60°. However they find it too difficult. Therefore the researcher observing them suggests that they mark a discrete trace, that is to mark the points corresponding to C as 60° (Figure 5)

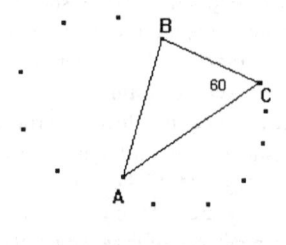

Figure 5

Cabri is then used to test this conjecture. The students do *soft locus dragging*, but they want to see the *soft locus*. They had been showed how to use 'trace'. Many students used it in the previous problem as a tool for exploration, in order to discover the path of C. In the case of testing a conjecture, however this tool is no more considered useful, as it gives a shaky trace, not the correct locus you are looking for. So the same tool is used by students in different ways according to different processes (discovering or testing). At this point the researcher suggests a new way of using Cabri, that reveals crucial for the students to prove their conjecture (that C is on a circle) false and to produce another one.

The students readjust their conjecture by saying that AB is a chord of 2 different circles. They also comment that "it doesn't work on the inside".

The final conjecture is produced from what they see in Cabri. The last remark from the students shows the attention on particular cases, and this statement can be worked on to lead students to formulate the complete theorem.

The two examples provide a window on the different ways of using dragging in Cabri and the cognitive processes with which they are associated. In particular a continuous shift from working in Cabri to students' reasoning and vice-versa can be observed. In some instances the students anticipate a conjecture and then they check it in Cabri, while in other cases they 'read' what is happening in Cabri and make conjectures from this. The *soft locus dragging* proved to be quite important in the discovering process, as it is the crucial point when students are able to make a conjecture, after the locus they 'feel' while moving in Cabri according to certain properties. Students seemed to exploit this tool spontaneously, however it is not always like this. For example the intervention of the researcher in the second problem, when he suggests the use of marking points satisfying a required property, shows the need to introduce some of the dragging tools or modalities in some way to students. These tools need to be made explicit and introduced into the classroom culture, so that they may become available to all students.

The evolution in the use of dragging is reflected also in the kind of conjectures students produced over time.

2 bisectors move at a time, other doesn't move. F never leaves the bisector which isn't moving.

When point A is moved, F moves along bisector BC.

When A is far away from the bisector BC, point F is outside the triangle.

These are examples of conjectures students wrote on the worksheet 'About triangles...' (session 4).

The first conjecture simply describes what moves and what does not move in the figure. The second one observes a fact that is due to the construction, it does not say anything new. The third conjecture is trying to say something more, a relation between two elements of the figure is discovered; the language used is 'when ... moves, ... is .../moves'. So there is a first stage, in which conjectures mainly describe the Cabri configuration or start to explore the relations between points. This seems to be common to all situations with students starting to use Cabri, as is supported by other observations.

If A is moved up or down the line AC then in effect you can make unlimited triangles, with C being the right angle.

If B is moved along the line BC then C will remain a right angle but the values of C and A will change.

If a circle is drawn, with its centre on the midpoint of AB. Then if C is moved to the perimeter of the circle then the angle ACB will be 90°.

In order for the triangle to be a right angle (where C is the right angle) C must be on the circle with diameter AB.

These are examples of conjectures students wrote on the worksheet *Angles in circles 1* (session 6).

In this second stage the language is different: all of the conjectures are formulated as logical sentences 'if ... then', however they are still embedded in the world of Cabri, in that they are expressed in terms of 'movement' or 'drawing'. They report the effects of moving different points (if A is moved, ...; if B is moved, ...) on the configuration.

If you have a triangle of 60° then the intersect of the perpendicular bisector of each side will be the centre point around which point C can be moved.

If you keep A and B fixed there are infinite 60° angles either side of the line AB. There are two curves linking A to B where 60° angles can be found.

C=60° on major segment or C=120° if in minor segment. To construct the diagram draw a normal circle and construct a triangle with all its points on the circle. B can be moved up any point on the line BC; A can be moved anywhere along the line AC and C will still be 60°.

These are examples of conjectures students wrote on the worksheet *Angles in circles 2* (session 8).

These conjectures are more elaborate. The different kind of dragging each point can undergo is stressed and the relationships between points are analysed in more detail (if you keep A and B fixed...). The second conjecture is the 'circle theorem' but formulated in the Cabri-language, that is in terms of fixed and movable points. So at this stage the logical form of the conjecture is still linked to what Hölzl (1996) refers to as the Cabri-geometry. The third conjecture contains the 'crystallisation' of the exploration process: it describes the way of constructing the diagram and of formulating the conjecture itself.

The main issue to be considered is the fact that by this time students actually produced rich conjectures that showed a certain appropriation of dragging. Students seemed to do experiments by moving different points and observing the related effect of dragging on the figure. They still use a Cabri language, as the examples above show. A next step would be that of 'translating' this Cabri-language into mathematical language, for example trying to transform the 'moving on the circle' into 'belonging to a circle'. Moreover the relationship between the way conjectures are formulated and the use of dragging still needs to be further investigated.

Other than producing conjectures, the evaluation questionnaire administered at the end of the project shows that students were developing the idea of what a conjecture is and what a proof is for. Examples of students' answers to the question 'what is a conjecture?' are:

"ideas that might be correct or incorrect that need to be proved"

"a theory or idea that you have come up with yourself and you believe to be true. It is not proved and may be found to be false if tested through"

"a prediction that hasn't been yet proved"

"a statement that is not proven yet and saying what you think is going to happen or what something is."

SOME CONCLUDING REMARKS

The first thing to be observed is the two-fold evolution of students' use of dragging in Cabri: within a session and over the whole intervention. It was observed that at the beginning the students did not use dragging in Cabri very much. This is a behaviour which has been observed in many experiments with students of different school levels just starting to use Cabri. They need to learn to move things around before they start doing this. Then an evolution of the use of dragging can usually be seen. My hypothesis to be investigated is that students need to 'internalise' (Vygotsky, 1978) the dragging function in order to be able to use it in a productive way. That is, at first dragging could even be seen as an element which distracts and interfere, since one is not used to see objects moving on paper; then one starts experimenting with dragging but, until a conscious use of it is reached, it is unlikely that students can really exploit it for what they want. Moreover, the use of different dragging modalities provides different approaches to the solution of a problem and different 'results' in terms of getting to a solution of the problem. These results suggest that these modalities might become objects of study among teachers and eventually object of teaching in the classroom, such that they become part of the classroom culture and a tool accessible to all students. The way of doing this still needs to be studied. In this respect, the role of the teacher becomes very important, as he/she is the one in charge of structuring didactical sequences such that the 'internalisation' takes place and becomes part of the classroom culture.

This learning sequence was structured in a way that alternated the Cabri sessions and classroom discussions or group work in the classroom. Besides, sometimes students felt the need for using other tools together with Cabri, such as sketches on paper or hand gestures (quite a lot of sketches were done on the worksheets). From these observations, it is likely that Cabri cannot provide a self-contained environment, but other tools and environments need to be used and studied in relation to Cabri. This issue has important didactical consequences, as it determines how Cabri is used in the classroom: 'Cabri and Cabri only', leaving responsibility for learning to the technology (Perkins, 1993), with the risk that no learning takes place, or integration of Cabri within a didactic process with specific learning objectives (in this case the experience was focused on using Cabri for conjecturing and towards proving in geometry). Again the role of the teacher appears crucial, as the need is that of structuring activities which allow students to choose which tools fit best.

The students' conjectures were still embedded in the Cabri geometry and used a Cabri language. This is what we might expect from students at the beginning and I would say this is the first step that can provide the basis for further development and evolution. Follow up questions might concern a way to start from what they did and provide a move towards justifying and towards the world of geometry. In what way can the Cabri-geometry evolve into geometry? How can the evolution of dragging modalities be a support towards this? How can the relational thinking, which relates different points through their connected movement and produces statements of the form "when I move this point then that other point moves", be transformed into functional thinking, that is into statements of the form "if...then", containing logical relationships between points?

This paper has focused on the process of conjecturing, however the process of proving needs to be studied with respect to the production of conjectures and the use of dragging. In what way can these conjectures provided the basis for proving? How would Cabri support the proving process, based on the conjecturing activity?

NOTES

1. I wish to acknowledge Mr. Jim Baker and his students for participating in this experience, and John Rogers for carrying out the project with me.

2. The research is funded by an ESRC research studentship.

3. AT1 is Attainment Target 1 of the National Curriculum. Children aged 11-14 years should be within the range of Levels 3 to 7. Level 8 is available for very able pupils.

4. Quotations are taken from the evaluation questionnaire, which was given to students at the end of the intervention.

5. The circle theorems are: 1. The angle at the circumference, standing on a diameter, is a right angle (angle in a semicircle); 2. Angles at the circumference, which stand on the same arc, are equal (angle in the same segment); 3. If an angle at the centre and an angle at the circumference stand on the same arc then the angle at the centre is double the angle at the circumference (angle at centre).

6. The original name given to this type of dragging was *lieu muet dragging,* which means dummy locus, as the locus is not visible and does not 'speak' to the students, who do not always realise that they are dragging along a locus.

7. Draggable points are basic free points, while semi-draggable points are points on objects.

8. Angles in circle 1: Draw any triangle ABC. Make it a right-angled triangle in C. How many right-angle triangles can you make? How can you characterise them? Make conjectures.

9. Angles in circle 2: Draw any triangle ABC. Now we want the triangle ABC to have the angle C of 60 degrees. How many triangles with this characteristic can you find? Which property characterises them?

10. The Cabri command 'trace' draws the path of a point while it is moved

APPENDIX: WORKSHEETS USED IN THE CLASSROOM

SESSION 3: GETTING STARTED WITH CABRI

Construct a segment AB. Draw the **midpoint** K of AB.

Drag A and B.

1. *What happens to K?*
2. *Can you move it?*

Draw another **point** H **on** AB.

3. *Can you move H? Where?*

Draw a point D outside AB. Construct the line s by D and **parallel** to AB.

4. *Can you move s? Why?*
5. *Drag A. What happens to the whole figure?*
6. *Drag D. What happens to the whole figure?*

Construct a right-angled triangle LMN.

Move L, M, N. Is the triangle still a right-angled triangle?

In Cabri we have a powerful instrument to check whether our construction is correct: it is called the **dragging test**. Drag any point of your figure:

if it keeps the property you want to then your construction is correct;

if it messes up then you need to redo your construction.

7. *Which points can you move in this construction? How?*

Construct an isosceles triangle and check it with the dragging test.

Construct an equilateral triangle and check it with the dragging test. Mark its angles and then measure them. Measure its sides.

8. *Move the vertexes and see what happens to the angles and the sides*

SESSION 4: ABOUT TRIANGLES

Construct a triangle ABC. Draw the perpendicular bisectors related to all its sides. Mark their intersection point and name it F.

1. *Move A. What changes and what stays the same in the figure?*
2. *Where is F with respect to the triangle?*
3. *Consider different configurations of the triangle (related to its sides and angles) and say where F is.*

Write down your conjectures.

Construct the circle centred on F and passing through A.

1. *What is the relationship between the triangle ABC and this circle?*

SESSION 6: ANGLES IN CIRCLE 1

Draw any triangle ABC.

Make it a right-angled triangle in C.

1. *How many right-angled triangles can you make?*
2. *How can you characterise them? Make conjectures.*
3. *Which points did you move?*

Explore the geometric situation in Cabri and write down your conjectures.

SESSION 8: ANGLES IN CIRCLE 2

Draw any triangle ABC.

Now we want the triangle ABC to have the angle C of 60 degrees.

1. *How many triangles with this characteristic can you find?*
2. *Which property characterises them?*

SESSION 9: ANGLES IN CIRCLES 3

STATEMENT 1:

IF AB is a diameter of a circle (centre O) and C is a point on the circle

THEN ....

Give a justification.

STATEMENT 2:

IF AB is a chord of a circle and C is a point on the circle

THEN ....

STATEMENT 3:

IF A, B and C lie on a circle and O is defined as the centre of the circle; comparing the angles <AOB and <ACB

THEN ....

REFERENCES

Arsac, G., Germain, G. and Mante, M.: 1988, *Probleme Ouvert et Situation-Probleme*, IREM, Academie de Lyon.

Arzarello, F., Olivero, F., Paola, D. and Robutti, O.: 1999, 'Dalle congetture alle dimostrazioni: una possibile continuità cognitiva', *L'Insegnamento della Matematica e delle Scienze Integrate*, *22B*(3), 209-233.

Arzarello, F. and Bartolini Bussi, M.: 1998, 'Italian trends in research in mathematical education: A national case study from an international perspective'. In A. Sierpinska and J. Kilpatrick (eds.), *Mathematics Education as a Research Domain: A search for identity* (pp. 197-212). Dordrecht: Kluwer Academic Publisher.

Arzarello, F., Gallino, G., Micheletti, C., Olivero, F., Paola, D. and Robutti, O.: 1998, 'Dragging in Cabri and modalities of transition from conjectures to proofs in geometry'. In A. Olivier and K. Newstead (eds.), *Proceedings of the 22nd Conference of the International Group for the Psychology of Mathematics Education* Vol. 2 (pp.32-39). Stellenbosch, South Africa.

Arzarello, F., Micheletti, C., Olivero, F., Paola, D. and Robutti, O.: 1998a, 'A model for analysing the transition to formal proofs in geometry.' In: A. Olivier and K. Newstead (eds.), *Proceedings of the 22nd Conference of the International Group for the Psychology of Mathematics Education* Vol. 2 (pp. 24-31). Stellenbosch, South Africa.

Balacheff, N.: 1998, *Apprendre la Preuve*. Grenoble: CNRS, Laboratoire Leibniz-IMAG.

Bartolini Bussi, M.: 1998, 'Joint activity in mathematics classrooms: A Vygotskian analysis'. In F. Seeger, J. Voigt and U. Waschescio (eds.), *The Culture of the Mathematics Classroom* (pp.1-45). Cambridge: Cambridge University Press.

Baulac, Y., Bellemain, F. and Laborde, J.M.: 1988, *Cabri-Géomètre, un Logiciel d'Aide à l'Apprentissage de la Géomètrie. Logiciel et manuel d'utilisation.* Paris: Cedic-Nathan.

Boero, P., Garuti, R. and Mariotti, M.A.: 1996, 'Some dynamic mental process underlying producing and proving conjectures.' In A. Gutierrez and L. Puig (eds.), *Proceedings of the 20th Conference of the International Group for the Psychology of Mathematics Education* Vol. 2 (pp. 121-128). Valencia.

Chazan, D.: 1993, 'High school geometry student's justification for their views of empirical evidence and mathematical proof.' *Educational Studies in Mathematics*, *24*, 359-387.

Duval, R.: 1991, 'Structure du raisonnement deductif et apprentissage de la démonstration.' *Educational Studies in Mathematics*, *22*, 233-261.

Goldenberg, E.P. and Cuoco, A.: 1998, 'What is dynamic geometry.' In R. Lehrer and D. Chazan (eds.) *Designing Learning Environments for Developing Understanding of Geometry and Space* (pp. 351-368). Hillsdale, NJ: Erlbaum.

Goldenberg, E.P.: 1995, 'Rumination about dynamic imagery.' In: R. Sutherland and J. Mason (eds.), *Exploiting Mental Imagery with Computers in Mathematics Education* (pp. 202-224). NATO Asi Series, Springer-Verlag.

Hanna, G.: 1996, 'The ongoing value of proof.' In A. Gutierrez and L. Puig (eds.), *Proceedings of the 20th Conference of the International Group for the Psychology of Mathematics Education* Vol. 1 (pp. 21-34).Valencia.

Harel, G. and Sowder, L.: 1996, 'Classifying processes of proving.' In A. Gutierrez and L. Puig (eds.), *Proceedings of the 20th Conference of the International Group for the Psychology of Mathematics Education* Vol. 3 (pp. 59-66). Valencia.

Healy, L. and Hoyles, C.: 1998, *Justifying and Proving in School Mathematics – Technical Report on the Nationwide Survey*. Institute of Education, University of London.

Hölzl, R.: 1996, 'How does "dragging" affect the learning of geometry.' *International Journal of Computers for Mathematical Learning, 1*, 169-187.

Hoyles, C.: 1997, 'The curricular shaping of student' approaches to proof', *For the Learning of Mathematics, 17*(1), 7-16.

Laborde, C.: 1998, 'Relationship between the spatial and theoretical in geometry: the role of computer dynamic representations in problem solving.' In D. Tinsley and D. Johnson (eds.), *Information and Communications Technologies in School Mathematics* (pp. 183-195). London: Chapman and Hall.

Laborde, C.: 1995, 'Designing tasks for learning geometry in a computer-based environment. The case of Cabri-Géomètre.' In: L. Burton and B. Jaworski (eds.) *Technology in Mathematics Teaching: A bridge between teaching and learning* (pp.35-68). Bromley: Chartwell-Bratt.

Mariotti, M.A., Bartolini Bussi, M., Boero, P., Ferri, F. and Garuti, R.: 1997, 'Approaching geometry theorems in contexts: from history and epistemology to cognition.' In E. Pehkonen (ed.) *Proceedings of the 21st Conference of the Internatoinal Group for the Psychology of Mathematics Education* Vol. 1 (pp. 180-195). Lahti.

Mariotti, M.A.: 1998, 'Introduzione alla dimostrazione all'inizio della scuola secondaria superiore.' *L'Insegnamento della Matematica e delle Scienze Integrate, 21B*(3), 209-252.

Mogetta, C., Olivero, F. and Jones, K.: 1999, 'Providing the Motivation to Prove in a Dynamic Geometry Environment.' In E. Bills (ed.) *Proceedings of the British Society for Research into Learning Mathematics Conference held at St. Martin's College, Lancaster* (pp.91-96). Coventry: University of Warwick.

Olivero, F.: 1998, *Un percorso didattico dalla congettura alla dimostrazione in geometria.* Unpublished degree dissertation. University of Turin.

Olivero, F: 1999, 'Cabri-géomètre as a mediator in the process of transition to proofs in open geometric situations.' *Proceedings of ICTMT 4 (International Conference for Technology in Mathematics Teaching).* Plymouth.

Pea, R.D.: 1987, 'Cognitive technologies for mathematics education.' In: A. Schoenfeld (ed.), *Cognitive Science and Mathematical Education* (pp.89-112). Hillsdale, N.J.: LEA.

Perkins, D.N.: 1993, 'Person-plus: a distributed view of thinking and learning.' In: G. Salomon (ed.), *Distributed Cognitions* (pp.88-110). Cambridge: Cambridge University Press.

Sutherland, R. and Balacheff, N.: 1999, 'Didactical complexity of computational environments for the learning of mathematics.' *International Journal of Computer for Mathematical learning, 4*, 1-26.

Usiskin, Z.: 1982, *Van Hiele Levels and Achievement in Secondary School Geometry. Final Report of the Cognitive Development and Achievement in Secondary School Geometry Project.* Department of Education, University of Chicago.

Vygotsky, L.S.: 1978, Mind in Society: The development of higher psychological processes. Harvard: Harvard University Press

NOTES ON CONTRIBUTORS

Carol Aubrey worked as a primary school teacher, researcher, educational psychologist and lecturer in education at universities in Cardiff and Durham. She is now Professor of Education and Director for the Centre for International Studies in Early Childhood at Christ Church University College Canterbury. During the last ten years she has been engaged in several research projects in the area of children's early development in mathematics and the processes of instruction.

Chris Bills is a Graduate Teaching Assistant in the Institute of Education, University of Warwick. He works with BA(QTS) and PGCE Secondary Mathematics students and is engaged with PhD research. He has previously taught mathematics in secondary schools and lectured in Mathematics Education at the University of Reading. His current research concerns the influence of the representations used by primary school teachers on the mental representations of their pupils.

Laurinda Brown is a lecturer in mathematics education at the University of Bristol, Graduate School of Education. A 14-year experience, in the same school, of teaching mathematics (in the later years being Head of Department) preceded a more varied experience as co-editor of the journal, *Mathematics Teaching* (for five years), curriculum developer and university mathematics education lecturer. Teaching mathematics remains a major interest and commitment and her research interests are based on and within such teaching. Her current interests are broadening into ways of knowing, especially seeing analytical and intuitive ways of knowing as non-opposed.

Soo Duck Chae is a PhD student in the Mathematics Education Research Centre at the Institute of Education, University of Warwick, where she is working on the construction of conceptual knowledge using experimental mathematics under the supervision of David Tall. She worked as a researcher in the field of graphic programming at the Korea Institute of Defence Analysis and as a lecturer at the Korea Military Academy and Seowon University. She enjoys fine art, especially oriental painting and computer graphics.

Alf Coles is currently head of mathematics at Kingsfield School, an 11-18 co-educational comprehensive school in South Gloucestershire. He taught for two years in Africa (one year in Zimbabwe and one in Eritrea) either side of completing an MA in Mathematics and Philosophy at Oxford University. His research interests include exploring the teaching and learning of algebra in year 7 classes (TTA Teacher Research Grant, ESRC funded project) and looking at teaching strategies using and supporting listening and hearing in secondary mathematics classrooms.

Cosette Crisan is a lecturer in pure mathematics at Newman College of Higher Education, Birmingham. She is studying for a PhD at South Bank University, London, and her research interest focuses on the interactions between mathematics teachers' use of the new technology and their professional knowledge base for teaching.

Sally Elliott is a PhD student based at Sheffield Hallam University. She spent two years teaching mathematics at a school in Sheffield before joining the PhD programme run jointly by Sheffield Hallam University and the University of Sheffield. Her current research interests include the use of graphical calculators and the role of student interaction in mathematical meaning making.

Derek Foxman taught mathematics in secondary schools in London for ten years and English in a kibbutz school in Israel. Subsequently he lectured in mathematics and then psychology in colleges of education before being appointed director of the APU monitoring team at the NFER in 1977. Other projects at the NFER included international studies of mathematics and science. He retired from the NFER as Head of Curriculum Studies in 1993 and since then he has worked as a freelance consultant in Britain and abroad. He has been a Visiting Lecturer at the University of North London and is currently a Visiting Fellow at the University of London Institute of Education.

Ray Godfrey taught mathematics in schools for fourteen years and has for sixteen years taught mathematics and mathematics education at Canterbury Christ Church University College, working mainly in initial training of primary teachers, including mathematics specialist. He has been involved in research projects investigating early mathematical development, educational deprivation and exclusion and subject knowledge of teaching students.

Tony Harries is a senior lecturer in mathematics education at the University of Durham. He has worked in the primary and secondary sectors of the education system both here and abroad. In addition to teaching and in-service work he has been involved in various research projects related to the learning of mathematics with particular reference to the development of numeracy competence and the use of Information and Communications Technology (ICT) in the learning of mathematics.

Brian Hudson is responsible for research and international development in the School of Education at Sheffield Hallam University. He is involved in teaching mainly on Masters and Doctoral programmes. His current research interests are in mathematics education, the use of Information and Communication Technology, open and flexible learning environments and comparative perspectives on the different traditions of Curriculum and Didaktik.

Keith Jones is a lecturer and researcher in mathematics education at the University of Southampton. Before taking up posts in Higher Education, he spent more than ten years teaching in predominantly multi-ethnic inner city comprehensive schools where he was actively involved in both curriculum development and classroom-based research. His areas of research expertise include the teaching and learning of geometry, the development of mathematical reasoning and proof, and the use of technology in mathematics teaching.

Candia Morgan is a senior lecturer in mathematics education at the Institute of Education, University of London, working with pre-service and in-service teachers.

Her current research interests include linguistics and language in mathematics education, teacher assessment, and the education of mathematicians. She is the author of *Writing Mathematically: The Discourse of Investigation*, Falmer Press.

Federica Olivero is a research student at the Graduate School of Education, University of Bristol. She graduated in mathematics from University of Turin, Italy, writing a final dissertation relating to mathematics education. Her current research interests focus on the mediating role played by new technologies in the teaching and learning of proof.

Declan O'Reilly is curriculum co-ordinator for mathematics education in the Division of Education at the University of Sheffield. His research interests include visualisation and the use of technology.

Pat Perks is a lecturer in mathematics education at the University of Birmingham, working with pre-service and in-service teachers. She taught secondary mathematics in Birmingham for twelve years, including eight as head of department, then became an advisory teacher in the same authority. Her current research interests include the place of ICT in the teaching of mathematics (particularly calculators, spreadsheets and the electronic whiteboard), teachers' subject knowledge, and the place of research in teachers' professional development.

Stephanie Prestage is a lecturer in mathematics education at the University of Birmingham working, with pre-service and in-service teachers. She spent a number of years teaching mathematics in London before moving Birmingham. She has a range of interests in mathematics teaching, mathematics education and professional development; with particular interests in teaching styles, role of teachers' subject knowledge; professional knowledge of teacher educators, implementation of policy into practice.

Tim Rowland is a lecturer in mathematics education at the University of Cambridge. Having tried and failed at doctoral research in axiomatic set theory, Tim taught mathematics in schools with marginally more success before moving to research in mathematics education and work with prospective and serving teachers. He is the author of *The Pragmatics of Mathematics Education*, Falmer Press. Recently, his research has focused on mathematics discourse analysis and on developing number-theoretic proofs through generic examples. He is currently collaborating in research on primary teachers' mathematics subject knowledge, together with colleagues in Cambridge, London, Durham and York.

Jim D. N. Smith is a senior lecturer in mathematics education at Sheffield Hallam University, working with pre-service and in-service teachers. He has 17 years previous experience as a teacher of mathematics. Research interests include initial teacher education and the teaching of mathematical modelling.

Rosamund Sutherland is Professor of Education at the Graduate School of Education, University of Bristol. Her research is concerned with the relationship between learning and the use of cultural tools such as textbooks, teachers and computer environments. Her early work focused on teaching and learning algebra with a particular emphasis on the role of computers.

David Tall is Professor in Mathematical Thinking at the Mathematics Education Research Centre, University of Warwick. He has previously held posts in primary mathematics at Woodloes Middle School, secondary science at Wellingborough Grammar School, university mathematics at Sussex and Warwick Universities, and mathematics research at The Princeton Institute for Advanced Study. He has authored or co-authored over 30 books and manuals, and over 200 papers on mathematics education, with special interests in computers, visualisation, symbolisation and cognitive growth in mathematics from child to adult. He enjoys the music of Percy Grainger, *The Bill*, and Ardbeg single malt.

Geoff Wake is a research fellow of the Centre for Mathematics Education at the University of Manchester. His work reflects interests he developed in teaching mathematics in schools and colleges. Currently his activities involve curriculum development and research across secondary and tertiary education with particular reference to post-16 mathematics where he has been working on curriculum design and research into the transfer of mathematics from educational settings to workplaces.

Jan Winter is a lecturer in education at the Graduate School of Education, University of Bristol Her research interests include assessment of mathematics and the professional development of teachers and she is convenor of the CLIO Centre for Assessment Studies in the Graduate School of Education. She was lead researcher on a project for SCAA on the transition from CGSE to A level mathematics. Her current research also includes work funded by the Nuffield Foundation looking at the implementation of a new programme of post-16 mathematics qualifications, an ESRC funded project on the development of algebra in Year 7 pupils and work for QCA on a publication for teachers on the formative uses of assessment in mathematics teaching.

Julian Williams is Professor of Mathematics Education, directing, researching and teaching in the Centre for Mathematics Education at the University of Manchester. His current enquiries include the study of learning, assessment and teaching of mathematics in schools and situated cognition and the transfer of mathematics from school/college settings to workplaces. A current interest in dialogue springs from its significance to both research methodology and teaching methods.